Aerosol Optics
Light Absorption and Scattering by Particles in the Atmosphere

Dr Alexander A. Kokhanovsky

Aerosol Optics

Light Absorption and Scattering by Particles in the Atmosphere

 **Springer**

Published in association with
Praxis Publishing
Chichester, UK

Dr Alexander A. Kokhanovsky
Institute of Environmental Physics
University of Bremen
Bremen
Germany

SPRINGER–PRAXIS BOOKS IN ENVIRONMENTAL SCIENCES
SUBJECT *ADVISORY EDITOR*: John Mason B.Sc., M.Sc., Ph.D.
EDITORIAL *ADVISORY BOARD MEMBER*: Dr Alexander A. Kokhanovsky, Ph.D. Institute of Environmental
Physics, University of Bremen, Bremen, Germany

ISBN 978-3-540-23734-1 Springer Berlin Heidelberg New York

Springer is part of Springer-Science + Business Media (springer.com)

Library of Congress Control Number: 2007935598

Cover design: Jim Wilkie
Project copy editor: Mike Shardlow
Author-generated LaTex, processed by EDV-Beratung, Germany

Printed on acid-free paper

Preface

The optical properties of atmospheric aerosol are of importance for a number of applications, including atmospheric visibility and climate change studies, atmospheric remote sensing and particulate matter monitoring from space. These applications are investigated at many research centers worldwide using spaceborne, airborne, shipborne, and ground-based measurements. Both passive and active instruments (e.g. lidars) are used. The primary interest lies in the determination of the vertical aerosol optical thickness, the single scattering albedo, the absorption and extinction coefficients, the phase function and the phase matrix. Vertical distributions of the aerosol properties are also studied using ground-based and spaceborne lidars. Considerable progress in understanding aerosol properties has been made in recent years. However, many problems still remain unsolved. They include, for instance, direct and indirect aerosol forcing, light interaction with nonspherical aerosol particles (e.g., desert dust), and also the retrieval of aerosol optical thickness and optical particle sizing using satellite observations.

The area of aerosol research is extensive. Therefore, no attempt has been made to achieve a comprehensive coverage of the results obtained in the area to date. The main focus of this book is the theoretical basis of the aerosol optics. The results presented are very general and can be applied in many particular cases. The first section is concerned with the classification of the different aerosol particles existing in the terrestrial atmosphere with respect to their chemical composition and their origin (e.g., dust and sea salt aerosols, smoke, and biological and organic aerosols). In the second chapter, I introduce the chief notions of aerosol optics, such as absorption, scattering, and extinction coefficients, and also phase functions and scattering matrices. Numerous examples of single scattering calculations using Mie theory are presented. Chapter 3 aims to describe techniques for the calculation of multiple scattering effects in aerosol media. The results are of importance for studies of light propagation in thick aerosol layers, where the single scattering approximation cannot be used. The discussion in this section is based on the solid ground of radiative transfer theory. Both scalar and vector versions of the theory are presented. Chapter 4 is focused on the Fourier optics of aerosol media. In particular, the reduction of contrast due to atmospheric effects and also the optical transfer functions of aerosol media are considered in detail. This section is of importance for understanding image transfer through the terrestrial atmosphere. The final chapter of the book is focused on the application of optical methods for the determination of aerosol microphysical and optical properties. Such topics as measurement of both direct and diffused solar light using Sun photometers and satellite remote sensing of atmospheric aerosol are covered. Also lidar measurements from ground and space are briefly touched upon in this chapter.

My hope is that this book will be useful to both students and engineers working in the area of aerosol optics and atmospheric remote sensing. I am grateful to the many colleagues who are invisible authors of this book. It is not possible to mention all of them in this preface but my special gratitude goes to Eleonora Zege for her encouragement during my first steps in science and also for shaping my approach to problem solving, to Vladimir Rozanov for his long-term collaboration in the area of radiative transfer, and to Wolfgang von Hoyningen-Huene and John Burrows for numerous discussions on the physical foundations of satellite remote sensing. I am also indebted to Clive Horwood, Publisher, for his encouragement, his patience, and his skill in the design and production of the book.

Alexander K. Kokhanovsky
Bremen, Germany
January 2008

Table of contents

Chapter 1. Microphysical parameters and chemical composition of atmospheric aerosol

1.1 Classification of aerosols

The optical properties of atmospheric aerosol are determined by chemical composition, concentration, size, shape, and internal structure of liquid and solid particles suspended in air. All these characteristics vary in space and time. At any time new particles can enter or leave the atmospheric volume under study. Also particles can be generated in this volume by gas-to-particle conversion processes. Very different particles are found in an elementary volume of atmospheric air. Depending on the aerosol type, one can identify among the particles different minerals, sulfates, nitrates, biological particles such as bacteria and pollen, organic particles, soot, sea salt, etc. These particles are very tiny objects with sizes typically around 100 nm. Therefore, usually they are not visible to the naked eye. Nevertheless, aerosol particles considerably reduce visibility, influence climate, and can cause health problems in humans.

There are three main sources of particulate matter in the terrestrial atmosphere. Particles can enter the atmosphere from the surface (e.g., dust and sea salt). Particles can be generated in the atmosphere by gas-to-particle conversions. Some the particles enter atmosphere from space (cosmic aerosol). Water and ice aerosols form clouds. They are treated in a separate branch of atmospheric science, namely, cloud physics. Clouds will be not considered here in a systematic way. Importantly, aerosol particles do not exist in isolation. They interact with cloud droplets, ice crystals, and gases. Also the interaction between aerosol particles (e.g., coagulation and coalescence) is of great importance for atmospheric science.

Surface-derived aerosol constitutes the main mass of suspended particulate matter with about 50 % contribution on a global scale. The particles born in the atmosphere dominate the aerosol number concentration. The cosmic aerosol influence is negligible in the lower atmosphere. However, it can influence atmospheric air properties in the higher atmospheric layers, where the concentration of terrestrial aerosol is low.

Humankind has important influences on a planetary scale. In particular, the concentration of trace gases increased considerably due to industrial activities and transportation. This is also the case for aerosols. At present the contribution of the anthropogenic aerosol to the total aerosol mass is significant (see Table 1.1). This leads to serious health problems in highly populated industrial areas. Also the anthropogenic aerosol is a major source of climate change. Greenhouse gases warm the planet and the anthropogenic aerosol acts in the opposite direction globally. Therefore, cleaning of the air in major cities with respect to suspended aerosol particles may lead to additional warming with respect to the current state.

Table 1.1. Emissions of main aerosol types. Reported ranges correspond to estimations of different authors (Landolt-Bornstein, 1988)

Aerosol type	Emission (10^6 tons per year)
Sea-salt aerosol	500–2000
Aerosol formed in atmosphere from a gaseous phase	345–2080
Dust aerosol	7–1800
Biological aerosol	80
Smoke from forest fires	5–150
Volcanic aerosol	4–90
Anthropogenic aerosol	181–396

For a correct simulation of light propagation in atmosphere, one needs to know the microphysical properties and type of aerosol in the propagation channel. This is rarely known in advance. Therefore, a number of models have been proposed to characterize average microphysical characteristics of aerosol depending on the location and, therefore, on the proportion of various types of particles (e.g., desert and oceanic aerosol models).

It is of importance to have a classification of main aerosol types. Then these types can be used as building blocks for the development of microphysical and optical aerosol models.

Atmospheric aerosols are usually classified in terms of their origin and chemical composition. The main aerosol types are given in Table 1.1.

Sea-salt aerosol (SSA) originates from the oceanic surface due to wave breaking phenomena. The largest droplets fall close to their area of origin. Only the smallest aerosol particles with sizes from approximately 0.1 to 1 μm (e.g., those formed by the bursting of bubbles at the ocean surface) are of a primary importance to the large-scale atmospheric aerosol properties. These particles can exist in the atmosphere for a long time. They have been identified over continents as well.

The shape of sea-salt aerosol particles depends on the humidity. Cubic particles (see Fig. 1.1) are found at low humidity. This is due to the cubic structure of sodium chloride, NaCl, the main constitute of SSA. NaCl is easily dissolved in water. Therefore, cubic forms transform into spherical shapes in high-humidity conditions. We see that SSA is extremely dynamic with respect to the modification of its shape. It is difficult to construct the universal optical model of SSA because of the considerable influence of shapes on the processes of light interaction with particles. At least two optical models of SSA are needed (i.e., for low- and high-humidity conditions). Yet another problem is associated with the fact that sea salt is not distributed uniformly in the aerosol particle formed by the attraction of water molecules in the field of high humidity. The concentration of NaCl molecules is larger close to the center of a particle as compared to its periphery. This leads to the necessity to account for the inhomogeneity of a particle in theoretical studies of its optical characteristics. The models of radially inhomogeneous particles must be used in this case. It is known that the internal inhomogeneity of particles considerably influences their ability to scatter and absorb light. Unfortunately, there are computational problems related to the calculation of optical characteristics in the case of nonspherical inhomogeneous particles. This leads to the widespread use of the homogeneous sphere model of an aerosol

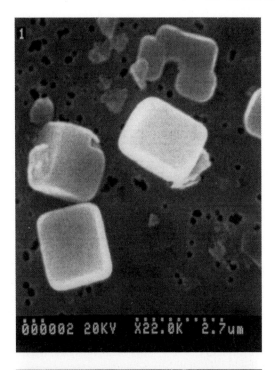

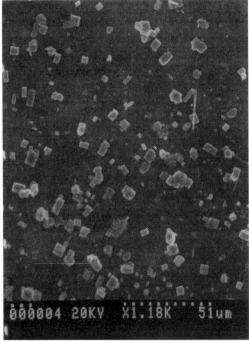

Fig. 1.1. Scanning electron photographs of dried sea-salt particles for marine air conditions collected at Mace Head on the west coast of Ireland (Chamaillard et al., 2003). The width of the picture represents 2.7 μm (top) and 51 μm (bottom).

particle. However, the danger is that this can potentially lead to a wrong interpretation of correspondent measurements.

The optical properties of aerosol particles are largely determined by the ratio a/λ, where λ is the wavelength of incident light and a is the characteristic size of a particle (e.g., the radius of a droplet or the side of a cubic crystal). Therefore, information on typical sizes of aerosol particles is of great importance for aerosol optics. This has been studied in numerous experiments. In particular, Clarke et al. (2003) found that dry sizes of sea-salt particles are in the range 0.1 to 10 μm. The number concentration N of sea-salt particles in the open ocean is usually around 250 cm^{-3}. The value of N is dominated by small particles with typical sizes around 0.3 μm. The particle size distribution (PSD) $f(a)$ of sea-salt aerosol is usually modeled using the lognormal law:

$$f(a) = \frac{1}{\sqrt{2\pi}\sigma a} \exp\left\{ -\frac{\ln^2(a/a_0)}{2\sigma^2} \right\}$$

normalized as

$$\int_0^\infty f(a)\, da = 1,$$

where a is the radius of a spherical homogeneous particle. It is often assumed for modeling purposes that $a_0 = 0.3$ μm and $\sigma = 0.9$. This means that the average radius of particles

$$\bar{a} = \int_0^\infty a f(a)\, da$$

and the coefficient of variance of PSD

$$C = \frac{\Delta}{\bar{a}}$$

with the standard deviation

$$\Delta = \sqrt{\int_0^\infty (a - \bar{a})^2 f(a)\, da}$$

are equal to 0.45 μm and 1.12, respectively, where we used relationships:

$$\bar{a} = a_0 \exp(0.5\sigma^2), \quad C = \sqrt{\exp(\sigma^2) - 1}.$$

The value of C is quite large (e.g., the standard deviation of PSD is equal to $1.12\bar{a}$). So the size distribution used is relatively broad.

Also the notions of the effective radius a_{ef} and the effective variance v_{ef} are used in aerosol optics. They are defined as follows:

$$a_{\text{ef}} = \frac{\int_0^\infty a^3 f(a)\, da}{\int_0^\infty a^2 f(a)\, da}, \quad v_{\text{ef}} = \frac{\int_0^\infty (a - a_{\text{ef}})^2 a^2 f(a)\, da}{a_{\text{ef}}^2 \int_0^\infty a^2 f(a)\, da}$$

or

$$a_{\text{ef}} = a_0 \exp(2.5\sigma^2), \quad v_{\text{ef}} = \exp(\sigma^2) - 1$$

for the lognormal PSD. It follows that $a_{\text{ef}} = 2.3$ μm and $v_{\text{ef}} = 1.25$ for the case of SSA considered above.

It is instructive to give the following relationships, which relate various forms of the particle size distribution representations:

$$f(a) \equiv \frac{dN}{da} = \frac{1}{a}\frac{dN}{d\ln a} = \frac{1}{a \ln 10}\frac{dN}{d\log a} = \frac{1}{\pi a^2}\frac{dS}{da} = \frac{3}{4\pi a^3}\frac{dV}{da} = \frac{3}{4\pi a^4}\frac{dV}{d\ln a} = \frac{3}{4\pi \rho a^3}\frac{dM}{da}.$$

Here V is the volume of particles, S is their geometrical cross section, $M = \rho V$ is the mass and ρ is the density of aerosol matter (2.2 g/cm^3 for sea salt). PSDs $f(a)$ are often characterized by the modal radius a_{m}, defined as the radius at which the distribution takes its maximum and, therefore, $df/da = 0$. Simple derivations enable us to obtain for the lognormal size distribution: $a_{\text{m}} = a_0 \exp(-\sigma^2)$, where the geometrical radius a_0 is the radius at which the following condition holds: $\ln a_0 = \overline{\ln a}$. Here the overbar means averaging with respect to PSD as defined above. In addition, it follows: $\sigma^2 = \overline{\ln^2(a/a_0)}$. In particular, one obtains: $a_{\text{m}} = 0.13$ μm at $a_0 = 0.3$ μm, $\sigma = 0.9$. A useful form of lognormal PSD is that at $\sigma = 1$. Then a number of relations are simplified. In particular, it follows in this case: $a_{\text{m}} = a_0/e$, where $e \approx 2.71828$. So the modal radius is about three times smaller as compared to the average geometrical radius a_0.

Recent optical measurements (Dubovik et al., 2002) suggest that the value of $\sigma = 1$ is at the upper border of the plausible change range of this parameter. In reality, the values of $\sigma = 0.4 - 0.8$ ($\ln\sigma \approx 1.5 - 2.2$) occur more frequently (for each mode of bi-modal PSDs). This means that the coefficient of variance C is in the range 0.4–1.0, with smaller values characteristic for the fine mode and larger values characteristic for the coarse mode. In the rough approximation, therefore, the standard deviation Δ of the coarse mode PSD is approximately equal to the average radius of aerosol particles in this mode and Δ is equal to the half of the average radius for the fine mode.

The interaction of optical waves with particles depends on the relative complex refractive index $m = n - i\chi$, which is close to 1.5 for sea salt in the visible with $\chi \approx 0$. However, one must account for the lowering of n due to humidity effects as discussed above (the refractive index of water is approximately 1.33 in the visible). Recommended values of m are given in Table 1.2 (WCP-112, 1986). It follows that the proposed value of n is close to 1.4 and the value of χ is small, suggesting that the absorption by oceanic aerosol can be neglected in the visible. The situation is different in infrared, however. Then χ starts to rise (see Table 1.2).

A comprehensive review of oceanic aerosol properties and dynamics was prepared by Lewis and Schwartz (2004). It contains about 1800 references on the subject and is a valuable source of information on marine aerosols.

Dust aerosol (DA) originates from the land surface. It is composed of solid particles. Most of particles (e.g., composed of Si) are not soluble in water. Therefore, dramatic changes of the aerosol particle shape and structure in the humidity field are rare events as compared to sea-salt aerosols. However, the mineral core can be covered by a water or ice shell in high-humidity conditions. This will modify the optical properties of the particle. Wet solid par-

Table 1.2. The refractive index $m = n - i\chi$ of oceanic, water-insoluble (mainly, dust), water-soluble (sulfates, nitrates, etc.), and soot aerosol, respectively (WCP-112, 1986)

λ, nm	n	χ	n	χ	n	χ	n	χ
300	1.40	$5.8 \cdot 10^{-7}$	1.53	$8.0 \cdot 10^{-3}$	1.53	$8.0 \cdot 10^{-3}$	1.74	0.47
400	1.39	$9.9 \cdot 10^{-9}$	1.53	$8.0 \cdot 10^{-3}$	1.53	$5.0 \cdot 10^{-3}$	1.75	0.46
550	1.38	$4.3 \cdot 10^{-9}$	1.53	$8.0 \cdot 10^{-3}$	1.53	$6.0 \cdot 10^{-3}$	1.75	0.44
694	1.38	$5.0 \cdot 10^{-8}$	1.53	$8.0 \cdot 10^{-3}$	1.53	$7.0 \cdot 10^{-3}$	1.75	0.43
860	1.37	$1.1 \cdot 10^{-6}$	1.52	$8.0 \cdot 10^{-3}$	1.52	$1.2 \cdot 10^{-2}$	1.75	0.43
1060	1.37	$6.0 \cdot 10^{-5}$	1.52	$8.0 \cdot 10^{-3}$	1.52	$1.7 \cdot 10^{-2}$	1.75	0.44
1300	1.37	$1.4 \cdot 10^{-4}$	1.46	$8.0 \cdot 10^{-3}$	1.51	$2.0 \cdot 10^{-2}$	1.76	0.45
1800	1.35	$3.1 \cdot 10^{-4}$	1.33	$8.0 \cdot 10^{-3}$	1.46	$1.7 \cdot 10^{-2}$	1.79	0.48
2000	1.35	$1.1 \cdot 10^{-3}$	1.26	$8.0 \cdot 10^{-3}$	1.42	$8.0 \cdot 10^{-3}$	1.80	0.49
2500	1.31	$2.4 \cdot 10^{-3}$	1.18	$9.0 \cdot 10^{-3}$	1.42	$1.2 \cdot 10^{-2}$	1.83	0.51

ticles generally have lower refractive indices as compared to particles in dry conditions. Therefore, humidity effects cannot be completely neglected. In contrast to the case of sea-salt aerosols, the nonsphericity of dust particles must be accounted for. This makes it difficult to model optical properties of dust aerosol. The numerical solution of the electromagnetic scattering problem can be obtained for many specific shapes of particles. The main problem is that it is not clear at this stage how to account for the diversity of shapes in a given dust-aerosol cloud, where almost every particle has a unique shape (see Fig. 1.2). This complicates theoretical studies of dust aerosol as compared to the case of spherical scatterers. Some calculations can be performed in the framework of discrete dipole or geometrical optics approximations (Kokhanovsky, 2004a). However, this is possible only for the limited spectral range. The successful attempt to bridge geometrical optics and exact T-matrix computations for spheroidal randomly oriented particles was reported by Dubovik et al. (2006). The corresponding database can be used to model optical characteristics of dust aerosols.

Experimental measurements of dust optical properties in the laboratory are also difficult (Muñoz et al., 2001; Volten et al., 2001; see http://www.astro.uva.nl/scatter). Moreover, particles for such measurements are collected on the surface. Therefore, they are representative only for near-surface conditions. Clearly, the sizes of dust particles are vertically distributed with smaller sizes and less dense materials dominating at larger altitudes. Also shape distributions vary with the distance from the ground surface. These effects are not accounted for in aerosol remote sensing at the moment. Okada et al. (2001) studied atmospheric mineral particles having sizes of 100–6000 nm using electron microscopy for samples collected in three arid regions of China. In all three regions, the mineral particles showed irregular shapes with a median aspect ratio v (ratio of the longest dimension to the orthogonal width) of 1.4 independently of the size. The ratio of the surface area to the periphery length or the circularity factor c was in the range 0.6–0.8 with smaller values for larger particles. The most probable value of c was close to 0.7. Note that it follows that $c = v = 1.0$ for spheres by definition. The most probable width-to-height ratio was in the range 2.0–5.0 suggesting that particles were mostly plate-like (see also Fig. 1.2).

Fig. 1.2. Scanning electron photographs of dust particles (Kalashnikova and Sokolik, 2004).

Apart from problems with the shape characterization of dust particles, there is a difficulty in assessing the refractive index of particles due to their complex internal composition (e.g., the mixture of different minerals as often seen in the grains of road dust). The homogenization techniques used often have no solid theoretical grounds. For instance, let us imagine that a mineral particle is composed of two substances, which are not internally mixed. Then, clearly, the optical properties of such a particle will differ from that calculated assuming an internal homogeneous mixture of minerals. There is also a problem with estimations of the imaginary part χ of the refractive index of dust aerosols. Clearly, χ strongly depends on the aerosol source (e.g., black, red or white soil, etc.).

At the moment very crude models of the microphysical properties of dust aerosols are used in optical modeling. In particular, it is often assumed that particles are spheres and are characterized by the lognormal size distribution with $a_{ef} = 10\,\mu m$, $C = 1.5$, $a_0 = 0.5\,\mu m$, $\sigma = 1.1$ (see the definitions of these parameters above). The value of the real part of the refractive index of dust is close to 1.5 in the visible (e.g., 1.53 at 550 nm) and decreases in the near-IR. The value of χ is often assumed to be equal to 0.008 in the visible and near-IR. It must be remembered, however, that these parameters may change considerably due to different aerosol source locations. Some data on the dust aerosol refractive index are summarized in Table 1.2. Comprehensive databases of aerosol refractive indices can be found at http://irina.eas.gatech.edu/data-ref-ind.htm and also at http://www.astro.spbu.ru/staff/ilin2/ilin.html.

The concentration N of aerosol dust particles varies considerably, depending on the aerosol mode. Hess et al. (1998) introduced three modes of mineral desert dust: the log-normal aerosol size distribution with $a_0 = 0.07, 0.39, 1.9$ µm and $\sigma = 0.67, 0.69, 0.77$, with larger values of parameters σ for larger a_0. They also proposed to use the cutoff radius of 7.5 µm, assuming that larger particles exist only close to the surface for large wind speeds. Therefore, these large particles do not influence the global optical dust characteristics. Hess et al. (1998) proposed the following values of N for these three modes: 269.5, 30.5, and 0.142 cm^{-3} with smaller values of N for larger a_0. Mineral dust is the heaviest among all aerosol types with a density of about 2.6 g/cm^3 (Hess at al., 1998). Further information on dust aerosols with comprehensive tables of their optical characteristics is given by d'Almeida et al. (1991) (see also the database located at the following website: http://www.lrz-muenchen.de/~uh234an/www/radaer/opac.html).

Secondary aerosol (SA) originates in the atmosphere due to gas-to-particle conversion. This aerosol is composed of mostly sulfates and nitrates. Also various organic substances (originating, for example, from gases emitted by plants) can make a large contribution in the total aerosol mass (Seinfeld and Pandis, 1998). In particular, SO_2 is oxidized to H_2SO_4 and the rate of conversion is influenced by the presence of heavy metal ions (e.g., Fe, Mn, V). Some of proposed reactions are given below:

$$SO_3 + H_2O \rightleftharpoons H_2SO_4,$$

$$NO + O_2 + H \rightarrow HNO_3,$$

$$SO_2 + C_2H_2 + allene \rightarrow C_3H_4S_2O_3,$$

$$NO_2 + hydrocarbons + photochemistry \rightarrow organic\ nitrates,$$

$$SO_2 + alkanes \rightarrow sulfinilic\ acid,$$

$$O_3 + olefines \rightarrow organic\ particles.$$

The generated particles are mostly of spherical shape with parameters of the lognormal distribution as follows: $a_0 \approx 0.1$ µm and $\sigma \approx 0.7$. The concentration N is usually in the range 3000–7000 cm^{-3}. The concentration can reach 15 000 cm^{-3} and even above this value for heavy pollution events (e.g., due to enhanced anthropogenic gaseous emissions).

The parameters mentioned above can vary considerably depending on the humidity. This aerosol is found at all locations. Therefore, it plays an important role in the global aerosol budget (Lacis and Mischenko, 1995). The modeled values of the refractive index of secondary water-soluble aerosol are given in Table 1.2.

The mixture of secondary aerosols with those generated at the surface (SSA, DA) at various concentrations can explain most of optical phenomena related to the propagation of light in the cloudless atmosphere. Peterson and Junge (1971) estimate that about 780 000 000 tons of secondary aerosol is produced every year with almost the same number for the combined surface derived aerosol (500 million tons of sea salt and 250 million tons of dust). This gives approximately $1.5 \cdot 10^9$ tons per year. Assuming, the density of aerosol particles of 1.5 g cm^{-3}, we find that the combined aerosol produced per year (if pressed into a solid cube) would occupy about 1 km^3. Adding to that anthropogenic emissions (about 10 %), forest fires, volcanoes and other aerosol sources will increase the volume of the cube even further. Although 1 km^3 is a small number on a planetary scale, it is the dispersion of aerosols in well-separated tiny particles, which makes aerosol so important for mass, heat, and energy transfer in the terrestrial atmosphere.

Let us consider now minor aerosol components. They give usually a minor contribution with respect to the total mass of suspended atmospheric particulate matter. However, these aerosols can play an important role on both the regional and the planetary scales. They can also make more than 50 % of the aerosol mass for short periods of time (e.g., pollen explosion events) or for selected locations.

Biological aerosols (BA) are characterized by the extreme particle size range and enormous heterogeneity. Biological material is present in the atmosphere in the form of pollens, fungal spores, bacteria, viruses, insects, fragments of plants and animals, etc. The volumetric concentration of bioaerosols depends on the season, location, and height of the sampling volume with smaller values at higher altitudes and in winter time (e.g., at high latitudes). Bioaerosols can occupy up to 30 % of the total atmospheric aerosol volume at a given location (especially in remote continental areas). Their concentrations are at least three times smaller in remote marine environments. Nevertheless, bioaerosols produced inland can travel very long distances owing to their low density. Darwin (1845) in his *The Voyage of the Beagle* describes the presence of various biological matter (mostly of inland origin) in the brown-colored fine dust which fell on a vessel. He was surprised to find stones with sizes of about 1 mm on the vessel. These stones and also much finer dust injured the astronomical instruments on the *Beagle*. Clearly, the impact of inland biological matter and dust is of a great importance for oceanic life-forms. Pollen has been collected thousands kilometers from its origin. This is used also for the identification of the origin of airmass at a given location. Spores of a number of molds were identified at 11 km height in the atmosphere (Cadle, 1966). Therefore, bioaerosols are widespread and occupy (in different proportions) the whole troposphere. Typical sizes of selected biological species are given in Table 1.3. It follows that the sizes of viruses and bacteria are quite small. This allows for their easy penetration of the respiratory systems of animals and humans. The size of pollen is larger and governed by other biological functions. Bioaerosols (e.g., viruses and bacteria) can be attached to other particles (e.g., dust, pollen, spores) and travel large distances using other particles, including cloud droplets, as a means of transportation.

Radii as given in Table 1.3 correspond to the cross-sectional area equivalent spherical particles. As a matter of fact, most biological aerosols are of nonspherical shape. In par-

Table 1.3. Sizes of biological particles

Biological particles	Radius, μm
Viruses	0.05–0.15
Bacteria	0.1–4.0
Fungal spores	0.5–15.0
Pollen	10.0–30.0

ticular, many bacteria are rod-shaped and cannot be characterized by just one size. Also bacteria have internal structure and cannot be considered as homogeneous objects in light scattering studies. Wittmaack et al. (2005) give a number of bioparticles as seen using scanning electron microscopy. Some images are shown in Fig. 1.3. It follows that particles have quite complex and variable shapes and internal structure. This makes it difficult to simulate the optical characteristics of such scatterers, even using advanced computers. Also the refractive index of particles needed for the theoretical modeling is poorly known. Due to the fact that in nature only left-handed amino acids and right-handed sugars exist, it is clear that most biological aerosols are chiral. This means that the refractive index depends on the sense of rotation of incident electromagnetic circularly polarized waves (Kokhanovsky, 2003). This issue must attract much more attention in future research due to its possible use in the problem of the remote detection of bioaerosols. It is emphasized that many biological particles not only absorb and scatter light: they can fluoresce when zapped with a beam of ultraviolet light. Jaenicke (2005) estimates the strength of the 'source biosphere' for atmospheric primary particles as equal to approximately 1000 Tg/year compared to 2000 Tg/year for mineral dust and 3300 Tg/year for sea salt.

Smoke aerosols (SMA) originate due to forest, grass, and other types of fires. Fires produce around 5 000 000 tons of particulate matter per year (Petterson and Junge, 1971). This is a small number as compared to the load of other aerosols. However, it has important local effects (e.g., as a cause of human, animal, and plant diseases; the reduction of visibility; and the changing of the heat balance) and an effect on global climate due to generally larger values of light absorption by smoke aerosol (e.g., black carbon) as compared

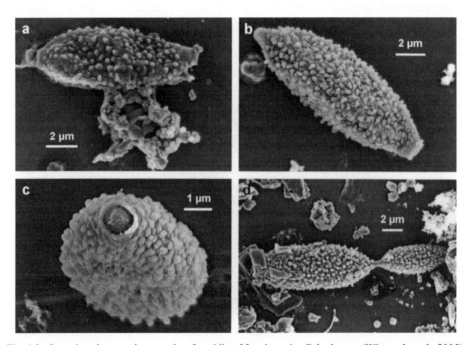

Fig. 1.3. Scanning electron photographs of conidia of fungi species C. herbarum (Wittmack et al., 2005).

to other aerosol species. The problem of black carbon influences on the planetary radiative budget is a hot topic in modern research. In particular, it has been found that aerosols transported to the Arctic from highly polluted areas in Europe can lead to a decrease in the planetary albedo (e.g, due to atmospheric absorption effects and also due to increased absorption of polluted snow and ice). Hansen and Nazarenko (2004) argued that dirty snow modifies planetary albedo and makes an important contribution to the global warming of the planet.

Smoke aerosols may lead to a number of spectacular optical atmospheric effects such as the blue Moon and Sun (van de Hulst, 1957).

Combustion processes produce tremendous numbers of small particles with radii below 0.1 μm. They also produce particles in the accumulation mode (0.1–1 μm), and 'giant' particles with radii above 1 μm. The number of particles with radii above 0.5 μm is relatively low. This means that particles of smoke can easily penetrate the respiratory system of humans leading to various health problems. Smoke aerosol has a large content of soot. Soot particles consist of aggregates with sizes generally greater than 1 μm in diameter with many particles of a smaller size as well. The aggregates are formed from the coalescence of ultimate (or primary) particles, which are in the range 50–100 nm.

Soot is often assembled in chain-like structures, which makes it impossible to use spherical particle models in estimations of their properties. The refractive index of soot varies, depending on its structure and production chain. Results for the soot refractive index shown in Table 1.2 can be used for optical modeling purposes, bearing in mind the great uncertainties especially with respect to the imaginary part of the soot refractive index.

Volcanic aerosols (VA) originate due to emissions of primary particles and gases (e.g., gaseous sulfur) by volcanic activity. Most of the particles ejected from volcanoes (dust and ash) are water-insoluble mineral particles, silicates, and metallic oxides such as SiO_2, Al_2O_3 and Fe_2O_3, which remain mostly in the troposphere. The estimated dust flux is 30 Tg per year. This estimate represents continuous eruptive activity, and is about two orders of magnitude smaller than dust emission. Volcanic sources can be important for the sulfate aerosol burden changes in the upper troposphere, where they might act as condensation nuclei for ice particles and thus represent a potential for a large indirect radiative forcing. Support for this contention lies in evidence of cirrus cloud formation from volcanic aerosols and some data that links the inter-annual variability of high-level clouds with explosive volcanoes. Volcanic eruptions can have a large impact on stratospheric aerosol loads. Volcanic emissions sufficiently cataclysmic to penetrate the stratosphere are rare. The stratospheric lifetime of coarse particles (dust and ash) is only about 1–2 months due to efficient removal by settling. Nevertheless, the associated transient climatic effects are large and trends in the frequency of volcanic eruptions could lead to important trends in average surface temperature. Sulfur emissions from volcanoes have a longer-lived effect on stratospheric aerosol loads. They occur mainly in the form of SO_2, even though other sulfur species may be present in the volcanic plume, predominantly SO_4^{2-} aerosols and H_2S. It has been estimated that the amount of SO_4^{2-} and H_2S is commonly less than 1 % of the total, although it may in some cases reach 10–20 %. Nevertheless, H_2S oxidizes to SO_2 in about 2 days in the troposphere or 10 days in the stratosphere. Estimates of the emission of sulfur-containing species from quiescent degassing and eruptions range from 7 to 14 Tg of sulfur per year. These estimates are highly uncertain because only very few of the potential sources have ever been measured and the variability between sources and between different stages of activity of the sources is considerable. The observed sulfate load in the stratosphere is about 0.14 Tg of S during volcanically quiet periods. The historical record of SO_2 emissions by erupting volcanoes shows that over 100 Tg of SO_2 can be emitted in a single event, such as the Tambora volcano eruption of 1815. Calculations with global climate models suggest that the radiative effect of volcanic sulfate is only slightly smaller than that of anthropogenic sulfate, even though the anthropogenic SO_2 source strength is about five times larger. The main reason is that SO_2 is released from volcanoes at higher altitudes and has a longer residence-time than anthropogenic sulfate.

Sulfate aerosol leads generally to cooling of a climate system. Therefore, there are theoretical studies with respect to the possibility of artificial production of this aerosol in the stratosphere with the aim of slowing down the current global warming trends (Crutzen, 2006). The pros and contras of such efforts must be carefully discussed before we enter the domain of artificial modification of climate.

Anthropogenic aerosol (AA) consists of both primary particles (e.g., diesel exhaust and dust) and secondary particles formed from gaseous anthropogenic emissions. Secondary liquid particles are quite small and their shapes can be approximated by spheres. However, a great portion of the anthropogenic aerosol mass is represented by irregularly shaped large particles, as shown in Fig. 1.4.

Anthropogenic aerosols contribute about 10 % of the total aerosol loading. However, these emissions did not occur in the pre-human era. The influence of this small (but growing) contribution on the climate system is not exactly known and must be assessed in future research (e.g., using studies of the Greenland and the Antarctic ice at different depths). Tsigaridis et al. (2006) used the following emissions in their coupled aerosol and gas-phase chemistry transport model:

- black carbon: 7.5 (2.1) Tg carbon per year,
- anthropogenic SO_2: 73 (2.4) Tg sulfur per year,
- carbone oxide: 1052 (219) Tg carbon per year,
- NO_x: 45 (9) Tg nitrogen per year,
- NH_4: 44 (7) Tg per year,
- volatile organic compounds (VOC): 251 (97) Tg carbon per year,
- primary organic aerosols: 44 (29) Tg per year,
- CH_2O: 19 (2) Tg per year,
- aromatic VOC: 14 (0) Tg per year,

where the numbers in parentheses give the preindustrial emissions (year 1860) of the corresponding substances. These numbers, although small in comparison with dust (1704 Tg/y) and especially sea-salt (7804 Tg/y) emissions, clearly indicate the influence of humankind on current atmospheric composition. The change in trace gases can have negative consequences leading to the destabilization of the climate system, to global warming, and to nonreversible processes in the Earth–atmosphere system. Therefore, it is of importance to monitor the trace gas vertical columns and also vertical concentrations of aerosols on a global scale using satellite measurements. It should be emphasized that, although the increased gaseous concentrations lead to larger atmospheric absorption and, therefore, to the warming of the atmospheric system, aerosol could increase or decrease the level of light reflection by the Earth–atmosphere system, depending on the ground albedo. For the low albedo typical for ocean and dark vegetation in the visible, additional anthropogenic aerosol leads to the increase of planetary albedo and, therefore, to cooling. Aerosol with a high soot content appears dark and, therefore, reduces planetary albedo, e.g., over bright snow surfaces. Additional warming can occur due to current efforts to clean polluted areas (e.g., cities in western Europe). This reduces the pollution load but increases risks related to heat waves (Meehl and Tibaldi, 2004).

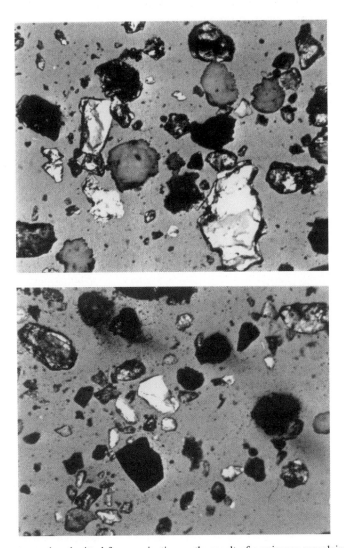

Fig. 1.4. A street sample submitted for examination as the result of a nuisance complaint. The magnification is equal to 40. The white snowballs are spheres of sodium carbonate from a nearby paper plant. In addition, the sample contains dried leaves, glass, glass fibers, paper fibers, cement dust, hematite, limestone, olivine, coal dust, soot, and burned wood. There is a great deal of quartz, covered wholly or partially by asphalt (McCrone et al., 1967).

1.2 Aerosol models

It follows from the discussion above that the microphysical characteristics of aerosols change considerably depending on the aerosol type, the season, etc. However, for modeling purposes, it is of importance to have simplified models which capture the main microphysical characteristics of atmospheric aerosol in a correct way and offer a simple way for the optical modeling. One such model, based on the assumption of the sphericity of aerosol particles, is given by Hess et al. (1998). The main parameters of the model are shown in Table 1.4. It is proposed to consider the aerosol at a given location as a mixture of certain aerosol components with prescribed size distributions, as shown in Table 1.5 for the case of continental aerosol. In particular, the sea-salt aerosol is divided into two fractions: fine mode and coarse mode. The coarse mode is of importance only for rough oceanic surface conditions occurring at high wind speed. Far from the oceans and also over calm water, the fine mode prevails. The desert dust aerosol is composed of three fractions with the behavior of the fractions similar to that of sea salt. In addition, the nucleation mode with very small particles is added. Only the spherical model of scatterers is considered, which is very remote from reality for mineral aerosols and dry sea salt. All other aerosol particles are subdivided into two broad categories: water soluble aerosol (e.g., sulfates and

Table 1.4. Microphysical properties of atmospheric aerosol components in the dry state (Hess et al., 1998)

Component	r_0, μm	σ
Sea salt (accumulation mode)	0.209	2.03
Sea salt (coarse mode)	1.75	2.03
Desert dust (nucleation mode)	0.07	1.95
Desert dust (accumulation mode)	0.39	2.0
Desert dust (coarse mode)	1.9	2.15
Water-insoluble aerosol	0.471	2.51
Water-soluble aerosol	0.0212	2.24
Soot aerosol	0.0118	2.0

Table 1.5. Composition of continental aerosol types. Mass values are given for a relative humidity of 50% and for a cutoff diameter of 15 μm (Hess et al., 1998). N_i is the number concentration and M_i is the mass concentration of the i-th aerosol component, n_i and m_i are correspondent number and mass mixing ratios

Aerosol type	Components	N_i, cm^{-3}	M_i, μg/m^3	n_i	m_i
Continental clean	Water-soluble	2600	5.2	1.0	0.591
	Insoluable	0.15	3.6	0.000 577	0.409
Continental average	Water-soluble	7000	14.0	0.458	0.583
	Insoluble	0.4	9.5	0.000 261	0.396
	Soot	8300	0.5	0.542	0.021
Continental polluted	Water-soluble	15 700	31.4	0.314	0.658
	Insoluble	0.6	14.2	0.000 12	0.298
	Soot	34 300	2.1	0.686	0.044

Table 1.6. Parameters of lognormal distribution of atmospheric aerosol components (WCP-112, 1986)

Component	r_0, nm	σ
Fine-mode (FM) aerosol (water-soluble aerosol)	5	1.0936
Coarse-mode (CM) aerosol (water-insoluble aerosol)	500	1.0936
Oceanic aerosol (OA)	300	0.9211
Soot aerosol (SA)	11.8	0.6931

Table 1.7. Composition of aerosol types (WCP-112, 1986)

Aerosol type	Component (volumetric concentration)
Continental	FM (29%), CM (70%), SA (1%)
Maritime	FM (5%), OA (95%)
Urban	FM (61%), CM (17%), SA (22%)

nitrates) and water-insoluble aerosol (e.g., soil). In addition, the soot component is introduced. This component is used to represent absorbing black carbon. Soot only weakly influences light scattering in atmospheric air, but black carbon is of primary importance for light absorption processes, especially in urban areas.

The model of Hess et al. (1998) includes several additional atmospheric processes such as the change of humidity. It is, therefore, superior with respect to earlier models based on fewer aerosol types (see, for example, Whitby, 1978; Shettle and Fenn, 1979; WCP-112, 1986). One such model, which is still in use, is represented in Table 1.6. The corresponding size distributions are illustrated in Fig. 1.5 and the mixing ratios for continental, maritime, and urban models are presented in Table 1.7.

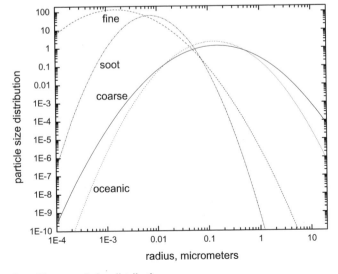

Fig. 1.5. Examples of lognormal size distributions.

Chapter 2. Optical properties of atmospheric aerosol

2.1 Introduction

Light can be scattered or absorbed by aerosol particles suspended in the terrestrial atmosphere. Processes of light scattering dominate over processes of absorption in the visible. However, absorption of light cannot be ignored. It influences the total radiation balance considerably. The reduction in the intensity of a direct beam during its propagation through an aerosol medium is determined simultaneously by absorption and scattering processes. The sum of total light scattering in all directions and absorption is called extinction. Energy, which is absorbed by particles, is not contained in them indefinitely, but rather radiates at larger wavelengths (emission). This book is mostly concerned with the effects of absorption and scattering. Emission, which is negligibly small at optical wavelengths, is neglected.

Light is composed of the superposition of electromagnetic waves having very high frequencies. To simplify, usually the idealized problem of a plane electromagnetic wave interaction with a single aerosol particle is considered under the assumption that an aerosol particle has a spherical shape. Then the electromagnetic field can be calculated both inside a particle (this is needed for the estimation of electromagnetic energy absorption effects) and at an arbitrary distance from a scatterer. The scattered energy can be integrated with respect to the direction of scattering yielding the scattering cross-section of a particle, which is defined as

$$C_{\text{sca}} = \frac{1}{I_0} \int_S I_{\text{sca}} \, dS,$$

where I_0 is the intensity of incident light and I_{sca} is the intensity of the scattered light, and S is the surface surrounding the particle. It follows from this definition that the scattering cross-section is measured in square meters. Usually, C_{sca} is smaller than the geometrical cross-section of an aerosol particle G defined as the projection of the particle on the plane perpendicular to the beam propagation direction. The ratio

$$Q_{\text{sca}} = \frac{C_{\text{sca}}}{G}$$

is called the scattering efficiency factor. One can also introduce the absorption efficiency factor

$$Q_{\text{abs}} = \frac{C_{\text{abs}}}{G},$$

where the absorption cross-section C_{abs} is defined in electromagnetic theory as

$$C_{abs} = \frac{k}{|\vec{E}_0|^2} \int_V |\vec{E}|^2 \varepsilon'' \, dV,$$

where V is the volume of a particle, $k = 2\pi/\lambda$, λ is the wavelength, $\vec{E}_0$ is the electric vector of the incident wave and $\vec{E}$ is the electric vector inside the scatterer. It follows that C_{abs} vanishes if the imaginary part of the dielectric permittivity $\varepsilon'' = 2n\chi$ is equal to zero. For most aerosol particles (with the exception of soot), the imaginary part χ of the refractive index of particles $m = n - i\chi$ is a small number in the visible (typically, smaller than 0.0001, see Table 1.2). This explains the relatively small absorption effects occurring during the interaction of light with aerosol media. The vector $\vec{E}$ and also scattered light intensity I_{sca} depend not only on the complex refractive index of a particle but also on its size, internal structure and shape. Mie theory (Mie, 1908) enables the quantification of size dependencies of various optical characteristics for the special case of spherical particles. Exact analytical solutions are also available for coated and multi-layered spheres (Kokhanovsky, 2004a).

2.2 Extinction

Although the chemical composition of aerosols and their size distributions are governed by a number of complex and not-well-understood processes, the resulting spectral aerosol extinction coefficient $k_{ext} = N\overline{C}_{ext}$, where N is the number of particles in a unit volume and $C_{ext} = C_{abs} + C_{sca}$ is the extinction cross-section (the overbar means averaging with respect to the aerosol size distribution), is often governed by the following simple analytical equation:

$$k_{ext} = b\lambda^{-a},$$

where a is called the Angstrom parameter and b gives the value of the aerosol extinction coefficient at the wavelength 1 μm, if the wavelength λ is expressed in micrometers. It follows from this formula that

$$\ln k_{ext} = \ln b - a \ln \lambda.$$

This provides a simple way to determine both a and b from experimental data (e.g., from spectral extinction measurements). The equations presented here are not exact ones and the actual dependence of k_{ext} on the wavelength can be a different one. In particular, nonlinear terms with respect to $\ln \lambda$ can appear, depending on the particular aerosol type. However, the dependence as given above closely represents average atmospheric conditions. Therefore, values of a and b are often measured. Long-term trends of the spectral exponent a are available at many worldwide locations. In particular, the Aerosol Robotic Network (AERONET) consists of more than one hundred identical globally distributed Sun- and sky-scanning ground-based automated radiometers (Holben et al., 1998). The network enables the determination of the Angstrom parameter and also a number of other aerosol characteristics, including size distributions, refractive indices and single-scattering albedos $\omega_0 = k_{sca}/k_{ext}$ (Dubovik et al., 2002). Here $k_{sca} = N\overline{C}_{sca}$ is the scattering coefficient.

AERONET measures not $k_{ext}(\lambda)$ itself but rather the spectral aerosol optical thickness (AOT):

$$\tau(\lambda) = \int_0^h k_{ext}(\lambda, z)\, dz,$$

where h is the top-of-atmosphere (TOA) altitude (e.g., 60 km) and z is the height above the ground level. The value of $t = \exp(-M\tau_t)$ gives the transmission coefficient of the direct solar beam flux after its propagation through the atmosphere. M is the air mass factor equal to the inverse cosine of the solar zenith angle μ_0 for the values of μ_0 not very close to zero. For a low Sun, corrections to the value of μ_0^{-1} taking into account the sphericity of atmosphere and also refraction must be taken into account. The value of τ_t also includes processes of gaseous absorption and molecular scattering (see the Appendix). Therefore, a special correction procedure is applied to remove these contributions and obtain AOT τ from the total atmospheric optical thickness τ_t.

Most aerosols are contained in the lower boundary layer with the height H (e.g., 1 km). Then, neglecting the vertical variation of the extinction coefficient, one derives:

$$\tau(\lambda) = k_{ext}(\lambda)H.$$

The Angstrom extinction law can also be presented in the following form:

$$\tau(\lambda) = \beta\lambda^{-a},$$

where $\beta \equiv \tau(1\ \mu m)$. Angstrom found that the value of a is close to 1.3 for the average continental aerosol (Angstrom, 1929). This has been confirmed by other researchers as well (Junge, 1963). The value of a can be related to the parameters of size distributions. Let us show this for the power law distribution:

$$f(a) = Aa^{-\gamma},$$

where A is the normalization constant and a is the radius of particles supposed to be bounded by the values a_1 and a_2. One can write for spherical particles:

$$k_{ext}(\lambda) = N \int_{a_1}^{a_2} \pi a^2 Q_{ext}(x, m(\lambda)) f(a)\, da,$$

where

$$Q_{ext} \equiv Q_{abs} + Q_{sca}$$

is the extinction efficiency factor, $x = 2\pi a/\lambda$ is the size parameter, and $m = n - i\chi$ is the refractive index. The value of Q_{ext} for different values of $n \in [1.01, 2.0]$ and $x \in [0.01, 100]$ is shown in Fig. 2.1. Oscillations seen in the figure are due to interference of diffracted and transmitted through a particle light. For large particles, $Q_{ext} \to 2$ as one might expect. The transition to this asymptotic limit is faster for larger values of n. For particles much smaller than the wavelength, Q_{ext} is a small number. Then particles do not influence light propagation in a great extent. The value of the extinction efficiency factor increases considerably in the resonance region ($a \sim \lambda$ and then approaches 2 for very large scatterers ($a \gg \lambda$). This corresponds to the so-called extinction paradox: the extinction cross-section is twice

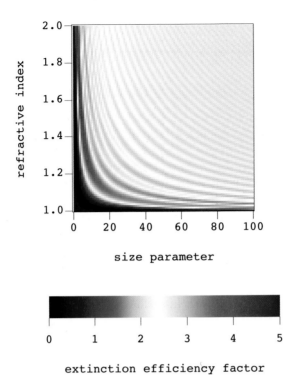

Fig. 2.1. Dependence of the extinction efficiency factor on the size parameter and refractive index of nonabsorbing particles calculated using Mie theory.

the geometrical cross-section of the particle. The paradox is solved, if one takes into account that not only rays incident on the particle but also those in the close vicinity of the scatterer are influenced by a particle due to the diffraction phenomenon.

The aerosol refractive index is a weak function of the wavelength in the visible. Neglecting its dependence on the wavelength and using the integration variable x instead of a, we derive for the power law distribution:

$$k_{ext}(\lambda) = B\lambda^{-(\gamma-3)} \int_{x_1}^{x_2} Q_{ext}(x) x^{2-\gamma}\, dx,$$

where $x_1 = 2\pi a/\lambda, x_2 = 2\pi a/\lambda, B = AN\pi(2\pi)^{\gamma-3}$. Clearly, the integral is a pure number. It does not depend on λ. So we conclude:

$$a = \gamma - 3.$$

Junge (1963) found from his measurements of aerosol PSDs that the average value of γ is close to 4.0. This gives: $a = 1$, which is in remarkable agreement with the Angstrom result $(a = 1.3)$ especially taking into account that completely different sets of measurements are involved in the derivation of a and γ. As a matter of fact, the relationship between a and parameters of the size distribution can be derived for any type of size distribution. Let us demonstrate this fact using the lognormal distribution introduced above and the refractive index of particles equal to 1.45 – 0.006i and the spectral exponent 1.53 + 0.008i.

We introduce the average extinction efficiency factor

$$\langle Q_{\text{ext}}\rangle = \frac{1}{S} \int\limits_{a_1}^{a_2} \pi a^2 Q_{\text{ext}}(a, \lambda) f(a)\, da,$$

where

$$S = \int\limits_{a_1}^{a_2} \pi a^2 f(a)\, da$$

is the average geometrical cross-section of spherical particles on the plane perpendicular to the beam. The dependence of $\langle Q_{\text{ext}}\rangle$ on the effective size parameter $x_{\text{ef}} = 2\pi a_{\text{ef}}/\lambda$ obtained using Mie theory for the lognormal size distribution with the coefficient of variance 1.0 is shown in Fig. 2.2. As expected, the extinction is stronger for particles with the larger refractive indices for the range of a_{ef} studied. It follows that $\langle Q_{\text{ext}}\rangle$ takes a maximum at the value of $x_{\text{ef}} \approx 10$ for the refractive indices used. If the largest contribution to the extinction comes from particles with sizes smaller than the wavelength of the incident light, then the extinction coefficient must decrease with the wavelength. The extinction efficiency factor $Q_{\text{ext}}(x)$ approaches its asymptotical value $Q_{\text{ext}} = 2$ as $x \to \infty$ from above (Kokhanovsky, 2006). This means that $\alpha = 0$ (no spectral dependence) for very large particles and α is negative for radii close to the asymptotic regime.

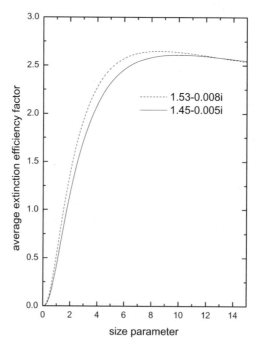

Fig. 2.2. Dependence of the extinction efficiency factor of spherical scatterers on the size parameter $x = ka$ at refractive indices 1.45–0.005i and 1.53–0.008i.

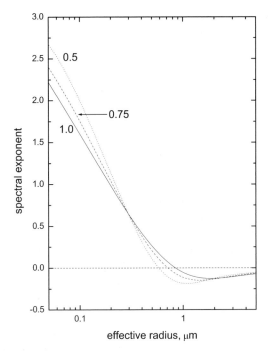

Fig. 2.3(a). Dependence of the spectral exponent on the effective radius of particles for the lognormal monomodal size distribution with the coefficient of variance equal to 0.5, 0.75, and 1.0. The refractive index $m = 1.45{-}0.005\mathrm{i}$ is assumed.

The dependence of the spectral exponent a on the effective radius of particles obtained using Mie theory and the integration as given above for a special case of lognormal size distribution with the coefficient of variance equal to 0.5, 0.75, and 1.0 is shown in Fig. 2.3(a). It follows that the derivation of a_{ef} from a is influenced by the coefficient of variance of the size distribution. The value of a was obtained using the following relationship:

$$a = c \ln(k_{\mathrm{ext}}(\lambda_1)/k_{\mathrm{ext}}(\lambda_2)),$$

where $c = \ln(\lambda_2/\lambda_1)$, $\lambda_1 = 0.412$ µm, $\lambda_2 = 0.67$ µm. Because the Angstrom law is only an approximation, the value of a depends on the pair of the wavelengths used. Therefore, reported values of a must identify the pairs of wavelengths used. Sometimes not just two wavelengths but rather the nonlinear fit of the complete spectral curve is used.

One can find from data shown in Fig. 2.3(a) that the dependence of a_{ef} (in µm) on a can be parameterized as follows:

$$\lg a_{\mathrm{ef}} = \sum_{j=0}^{4} d_j a^j,$$

where $d_0 = -0.07075$, $d_1 = -1.03109$, $d_2 = 0.72806$, $d_3 = -0.41111$, $d_4 = 0.08106$ at the coefficient of variance equal to 1. The accuracy of the fit for the curve $a_{\mathrm{ef}}(a)$ is demonstrated in Fig. 2.3(b).

The correct theoretical approach to the determination of a_{ef} and also $f(a)$ is the inversion of the integral

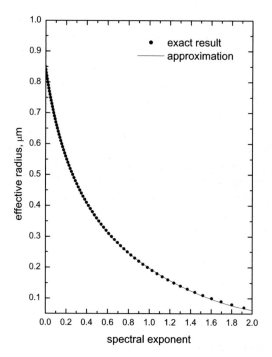

Fig. 2.3(b). Dependence of the effective radius on the spectral exponent calculated using exact Mie calculations and the approximation discussed in the text for the lognormal monomodal size distribution with the coefficient of variance equal to 1.0. The refractive index $m = 1.45 - 0.005i$ is assumed.

$$k_{ext}(\lambda) = N \int_{a_1}^{a_2} \pi a^2 Q_{ext}(a, \lambda) f(a) \, da$$

with respect to unknown function $f(a)$. For these special inversion techniques are applied (Twomey, 1977; Shifrin, 2003). Clearly, the inversion is not possible, if $Q_{ext} = 2$, which is a valid assumption for very large particles with $a \gg \lambda$ as, for instance, in the case of desert dust outbreaks. Then it follows that

$$k_{ext} = 2NS$$

independently of the shape, size distribution, and internal structure of particles. Here S is the average projection area of particles on the plane perpendicular to the incident beam. One can prove (Vouk, 1948) that S is equal to the quarter of the surface area Σ for convex randomly oriented particles. This means that $\varsigma \equiv N\Sigma = 2k_{ext}$ gives the total surface area of particles per a unit volume of an aerosol medium with particles having dimensions much larger than the wavelength. This is an important parameter for atmospheric chemistry. For instance, it determines the reactive surface for gas–aerosol interactions.

Ignatov and Stowe (2002) suggested that the frequency distribution of a for a given location follows a Gaussian law:

$$f(a) = \frac{1}{\sqrt{2\pi}\delta} \exp\left(-\frac{(a - a_m)^2}{2\delta^2}\right),$$

where a_m is the arithmetic mean and δ is the standard deviation of the Angstrom exponent. Smirnov et al. (2002) reported values of a equal to 0.76 and 0.93 for two remote sites in the Pacific Ocean (Lanai, 722 measurements) and the Atlantic Ocean (Bermuda, 590 measurements), respectively. These results together with measurements over continents suggest that $k_{ext} \sim 1/\lambda$ on average. However, deviations often occur. Atmospheric measurements show that a changes between -0.1 and 2.5 (Dubovik et al., 2002). The upper value of a is somewhat uncertain. This is due to the fact that large values of a often occur for optically thin aerosols with $\tau \ll 1$. The measurement of a at small values of τ is difficult due to errors of sun photometers, which are at least of the order of 0.01 in the value of the optical thickness. The largest possible value of a is close to 4. This takes place for a purely molecular atmosphere with scatterers having sizes $a \ll \lambda$. The values of $a = 3$ and above can be found in laboratory measurements. However, they are not characteristic for *in situ* atmospheric measurements.

The extinction coefficient depends not only on the size of particles but also on their concentrations. Its value is usually in the range 0.1–0.5 km^{-1} at 550 nm at urban locations (Horvath and Trier, 1993) and much lower for clean rural and remote oceanic regions. The meteorological range of visibility is calculated as

$$v = \frac{|\ln 0.02|}{k_{ext}(550 \text{ nm})} \approx \frac{3.91}{k_{ext}(550 \text{ nm})}.$$

So v usually ranges from 8 to 40 km for urban locations given by Horvath and Trier (1993).

2.3 Absorption

Extinction of light by aerosol is due to both light scattering and absorption. Usually light absorption is small and the ratio of absorption ($k_{abs} = N\overline{C}_{abs}$) to extinction ($k_{ext} = N\overline{C}_{ext}$) coefficients is smaller than 0.1 and even 0.01 for remote clean areas. The primary absorber of light in the atmosphere is soot (Horvath, 1993). Therefore, the probability of photon absorption $\beta = k_{abs}/k_{ext}$ differs from zero substantially only if soot is present in great amount. This is usually the case for urban areas. The value of the absorption coefficient of aerosol is of importance for climate change problems. It determines the cooling or warming effect of aerosol above a scene with a given ground albedo. Therefore, considerable efforts have been undertaken to characterize the spatial distribution of k_{abs} and also of single-scattering albedo $\omega_0 = 1 - \beta$. Some results in this direction are reported by Dubovik et al. (2002). Generally, the characterization of atmospheric aerosol absorption is a very complicated matter. Therefore, it comes as no surprise that there is a lot of uncertainty in our understanding of solar light absorption by atmospheric particulate matter.

It is much more difficult to measure the atmospheric aerosol absorption $k_{abs} = k_{ext} - k_{sca}$ as compared to the aerosol extinction. This is mostly due to the fact that the amount of energy absorbed by aerosols is much smaller as compared to the light scattered energy. There are a number of techniques to derive the aerosol absorption (see, for example, http://www.dfisica.ubi.pt/ $\sim$ smogo/investigacao/references.html). The most frequently used techniques are:
- filter methods;
- optical acoustic spectrometry;

- the diffuse transmittance method;
- techniques based on the measurement of differences between light extinction and scattering for a given atmospheric volume;
- the radiative transfer retrieval approach based on the measurements of the aerosol optical thickness, small-angle scattering, and diffuse light transmission by a sun photometer;
- polarimetric techniques.

Filter methods are based on studies of light absorption by measuring the optical attenuation through a filter on which aerosol particles have been accumulated. Measurements can be performed in real time during the sampling process using an aethalometer (Hansen et al., 1994). This measurement can take place after sampling has been terminated (e.g., integrating plate technique (Lin et al., 1973)). The shortcomings of filter techniques are numerous. In particular, the sampling process on the filter, the change in the shape of particles, and the interaction of scattering and absorption by densely aggregated aerosols must be well understood and accounted for in the derivation of the absorption coefficient, if possible. Another problem is due to the fact that the filter/plate transmission is not properly described by Beer's law. In addition, filter methods do not account for the influence of humidity on aerosol light absorption, as, even for real-time methods such as the aethalometer, the humidity-dependence of aerosol light absorption is likely to be modified by the filter substrate. As a consequence, integrating plate type measurements, for example, require the use of an empirical calibration factor.

Optical acoustic spectrometry is based on measurements of energy of a sound wave generated by the expansion of the air due to aerosol absorption of light from a modulated source. This absorption causes periodic heating and the subsequent expansion of the surrounding air at the modulation frequency (Pao, 1977). The practical use of the method is limited due to relatively low sensitivity and the use of inefficient light sources. However, important progress has been made in recent years to overcome the limitations of the technique (Lack, 2006).

The diffuse transmittance method is based on the measurements of the diffuse transmitted flux. The deviation from the radiative transfer model calculations for the case of $\omega_0 = 1$ enables the determination of aerosol single-scattering albedo (von Hoyningen-Huene et al., 1999).

Techniques based on the subtraction of scattering, for example, measured by a nephelometer, from extinction are valid only for the case of strong absorption. Otherwise, the value of the absorption coefficient is close to the level of noise and large errors in the measured difference are possible.

The radiative transfer retrieval approach is routinely used by AERONET. Single-scattering albedo is derived from simultaneous measurements of small-angle scattering, aerosol optical thickness, and diffuse light intensity in almucantar fitting measurements to the most probable atmospheric aerosol model. The technique is limited by *a priori* assumptions assumed in calculations. Dubovik et al. (2002) used this technique to determine typical values of the spectral single-scattering albedo at a number of locations worldwide. In particular, they give values of ω_0 equal to 0.78–0.88 for biomass burning in savanna (Zambia), 0.83–0.90 for Mexico City, 0.93–0.99 for the mixture of desert dust and oceanic aerosol in the Atlantic (Cape Verde). The first number corresponds to the wavelength 440 nm and the second one is measured at 1020 nm. The results give average values based on a great number of measurements (e.g., 91 500 measurements for Cape Verde (1993–2000)). It follows that the probability of photon absorption $\beta = 1 - \omega_0$ decreases with the wavelength and it is equal to 0.22 at 440 nm for biomass burning events in Zambia. It equals 0.17 for Mexico City and 0.07 for Cape Verde at the same wavelength. The value of β (440 nm) is just 0.02 at remote clean marine environments (Hawai, 1995–2000). The main absorbing component of the atmospheric aerosol in the visible is soot. Therefore, locations with small fractions of soot in atmospheric air are characterized by smaller light absorption levels.

Polarimetric techniques use an approach similar to that just described. However, not only intensity but also the polarization of scattered light is analyzed (Masuda et al., 1998; Kokhanovsky, 2003). The technique is highly sensitive to the refractive index of atmospheric aerosol. However, the retrievals are model-dependent and usually performed for the case of spherical scatterers. It is known that the polarization characteristics are highly influenced not only by the size and refractive index of particles but also by their shape and internal structure.

The great variety of existing techniques also suggests that there is no a single reliable technique for the measurement of atmospheric aerosol absorption in situ. This leads to great uncertainties in the estimations of aerosol radiative forcing.

On the other hand, aerosol absorption coefficient $k_{abs} = N\overline{C}_{abs}$ can be easily calculated in the case of spherical particles, if their size and refractive index are known. Such calculations are useful because they identify possible spectral dependencies to be found in correspondent measurements. For very small absorbing aerosol particles, the absorption cross-section is proportional to the volume of particles and also to the bulk absorption coefficient $a = 4\pi\chi/\lambda$ ($m = n - i\chi$ is the complex refractive index of particles relative to the surrounding medium):

$$C_{abs} = DaV,$$

where the value of $D \rightarrow 1$ as $n \rightarrow 1$ for arbitrary shapes of particles. It follows that (Shifrin, 1951):

$$D = \frac{9n}{(n^2 + 2 - \chi^2)^2 + 4n^2\chi^2}$$

and

$$D = \frac{9n}{(n^2 + 2)^2}$$

at $\chi \ll n$. This simple equation for C_{abs} follows from the definition of the absorption coefficient via the volume integral as presented at the beginning of this chapter. In derivations one must account for the fact that, as known from the electrodynamics, the electric field inside a spherical particle with the radius $a \ll \lambda$ (and also $k|m - 1|a \ll \lambda$) is given by the following simple equation:

$$\vec{E} = \frac{3}{m^2 + 2}\vec{E}_0.$$

It follows that internal and incident fields coincide $\left(\vec{E} = \vec{E}_0\right)$ at $m = 1$ as one might expect.

In the case of large spherical particles (under assumptions $x \to \infty, 2ax \to \infty$) all light, which penetrates the surface of a particle is absorbed. Then the absorption efficiency factor is equal to the fraction of light energy A, which penetrates the surface (Kokhanovsky, 2004a). This fraction can be estimated from the fraction of reflected energy: $A = 1 - r$. The value of r for very large particles can be found using Fresnel reflection coefficients R_i:

$$r = \frac{1}{2}\sum_{i=1}^{2}\int_{0}^{\pi/2}|R_i(\varphi)|^2\sin\varphi\cos\varphi\,d\varphi,$$

where

$$R_1 = \frac{\cos\varphi - m\cos\psi}{\cos\varphi + m\cos\psi}, \quad R_2 = \frac{m\cos\varphi - \cos\psi}{m\cos\varphi + \cos\psi},$$

φ is the incidence angle and $\psi = \arcsin(\sin\varphi/m)$ is the refraction angle. This integral can be evaluated analytically under assumption that $\chi \ll n$ and $n > 1$:

$$r = \frac{8n^4(n^4 + 1)}{(n^4 - 1)^2(n^2 + 1)}\ln n + \frac{n^2(n^2 - 1)^2}{(n^2 + 1)^3}\ln\left(\frac{n - 1}{n + 1}\right) + \frac{\sum_{l=0}^{7}p_l n^l}{3(n^4 - 1)(n^2 + 1)(n + 1)},$$

where $p_l = (-1, -1, -3, 7, -9, -13, -7, 3)$. The values of r are presented in Table 2.1 as a function of the refractive index n. For quick estimations, one can use the following parameterization: $r = 0.1396n - 0.1185$ valid for the range of refractive indices shown in Table 2.1.

Finally, we conclude that $C_{abs} = (1 - r)S$ at $x \gg 1$, $2ax \gg 1$, where the value of S is the geometrical cross-section of the particle. This formula is valid both for spherical and nonspherical large aerosol particles.

The absorption coefficient of air with inclusions of absorbing spherical particles characterized by the size distribution $f(a)$ can be calculated using Mie theory. It follows that

$$k_{abs}(\lambda) = N\int_{a_1}^{a_2}\pi a^2 Q_{abs}(a, \lambda)f(a)\,da,$$

Table 2.1. The dependence of the factor r on the refractive index n

n	r	n	r	n	r
1.333	0.0664	1.5	0.0918	1.7	0.1203
1.35	0.0691	1.55	0.0991	1.9	0.1475
1.4	0.0768	1.6	0.1063	2.0	0.1606
1.45	0.0844	1.65	0.1133	2.1	0.1734

where Q_{abs} is the absorption efficiency factor. The value of Q_{abs} depends on the spectral refractive index of aerosol particles and also on the size parameter $x = 2\pi a/\lambda$. The corresponding theoretical dependence is quite complicated (van de Hulst, 1957).

Namely, it follows that

$$Q_{abs} = Q_{ext} - Q_{sca},$$

where, according to Mie theory,

$$Q_{ext} = \frac{2}{x^2} \sum_{n=1}^{\infty} (2n+1)\,\mathrm{Re}(a_n + b_n),$$

$$Q_{sca} = \frac{2}{x^2} \sum_{n=1}^{\infty} (2n+1)\left[|a_n|^2 + |b_n|^2\right].$$

Here,

$$a_n = \frac{\psi'_n(y)\psi_n(x) - m\psi_n(y)\psi'_n(x)}{\psi'_n(y)\xi_n(x) - m\psi_n(y)\xi'_n(x)},$$

$$b_n = \frac{m\psi'_n(y)\psi_n(x) - \psi_n(y)\psi'_n(x)}{m\psi'_n(y)\xi_n(x) - \psi_n(y)\xi'_n(x)},$$

$m = n - i\chi$ is the relative refractive index of a particle ($m = m_p/m_h$, m_p and m_k are the refractive indices of a particle and a host medium respectively), $y = mx$, $x = \dfrac{2\pi a}{\lambda}$, a is the radius of a particle, λ is the incident wavelength in a host nonabsorbing medium, $\psi_n(x) = \sqrt{\dfrac{\pi x}{2}} J_{(n+1/2)}(x)$, $\xi_n(x) = \sqrt{\dfrac{\pi x}{2}} H^{(2)}_{n+(1/2)}(x)$, $J_{n+(1/2)})$ and $H^{(2)}_{n+(1/2)}$ are Bessel and Hankel functions.

For two-layered particles, the amplitude coefficients a_n and b_n take the following forms:

$$a_n = \frac{\psi_n(y)\left[\psi'_n(m_2 y) - A_n\chi'_n(m_2 y)\right] - m_2\psi'_n(y)\left[\psi_n(m_2 y) - A_n\chi_n(m_2 y)\right]}{\xi_n(y)\left[\psi'_n(m_2 y) - A_n\chi'_n(m_2 y)\right] - m_2\xi'_n(y)\left[\psi_n(m_2 y) - A_n\chi_n(m_2 y)\right]},$$

$$b_n = \frac{m_2\psi_n(y)\left[\psi'_n(m_2 y) - B_n\chi'_n(m_2 y)\right] - \psi'_n(y)\left[\psi_n(m_2 y) - B_n\chi_n(m_2 y)\right]}{m_2\xi_n(y)\left[\psi'_n(m_2 y) - B_n\chi'_n(m_2 y)\right] - \xi'_n(y)\left[\psi_n(m_2 y) - B_n\chi_n(m_2 y)\right]},$$

where

$$A_n = \frac{m_2\psi_n(m_2 x)\psi'_n(m_1 x) - m_1\psi'_n(m_2 x)\psi_n(m_1 x)}{m_2\chi_n(m_2 x)\psi'_n(m_1 x) - m_1\chi'_n(m_2 x)\psi_n(m_1 x)},$$

$$B_n = \frac{m_2\psi_n(m_1 x)\psi'_n(m_2 x) - m_1\psi_n(m_2 x)\psi'_n(m_1 x)}{m_2\chi'_n(m_2 x)\psi_n(m_1 x) - m_1\psi'_n(m_1 x)\chi_n(m_2 x)},$$

and m_1, m_2 are relative to a host medium refractive indices of a core and shell respectively, $x = ka$, $y = kb$, $k = 2\pi\lambda$, a is the radius of a core, b is the radius of a particle. Numerical calculations using equations shown above and also corresponding computer codes are discussed by Bohren and Huffman (1983) and also by Babenko et al. (2003).

The results of numerical calculations of Q_{abs} using Mie theory are shown in Fig. 2.4(a) for the same range of parameters n and x as in Fig. 2.1. It was assumed that the imaginary part of the refractive index is equal to 0.008. Single-scattering albedo calculated for the same conditions as in Fig. 2.4(a) is presented in Fig. 2.4(b). It follows that single-scattering albedo generally increases with the size of particles.

Often approximate relations for the local optical parameters of an aerosol medium are used. They make it possible to avoid tedious numerical calculations and to make quick estimates of corresponding optical parameters. Introducing the volumetric concentration of particles, $c = N\overline{V}$, one derives for absorbing particles with radii much smaller than the wavelength:

$$k_{abs}(\lambda) = cD(n)a(\lambda).$$

This expression differs from that for homogeneous media without scattering due to the presence of the coefficient D. This coefficient (see above) is equal to one at $n = 1$, as it should be. It is equal to 0.5 at $n = 2$. Therefore more reflective particles are less absorbing. For the typical value of $n = 1.7$ (soot), we have: $D = 0.7$.

The complication is due to the fact that very small soot grains with sizes of about 50 nm agglomerate in long chains. Then the dimension of this complex particle becomes too large and the approximation described above cannot be used. Also Mie theory cannot be applied

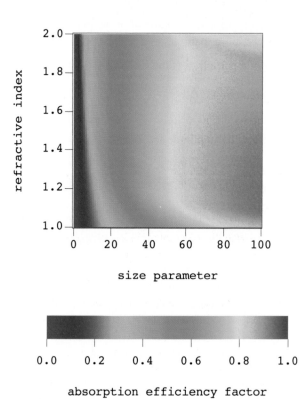

Fig. 2.4(a). Dependence of the absorption efficiency factor on the size parameter and refractive index of aerosol particles calculated using Mie theory at the imaginary part of the refractive index equal to 0.008.

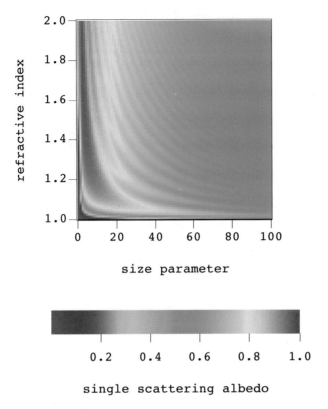

Fig. 2.4(b). Dependence of the single-scattering albedo on the size parameter and refractive index of aerosol particles calculated using Mie theory at the imaginary part of the refractive index equal to 0.008.

due to the nonsphericity of the resulting particle. There are indications (Berry and Percival, 1986) that aggregation will increase the absorption cross-section of a particle as compared to the case of the sum of cross-sections of isolated soot grains. Such an enhancement also occurs if soot grains are incorporated inside large nonabsorbing particles. This is due to the focusing effect of a nonabsorbing aerosol particle or a fog droplet. This means that the density of electromagnetic radiation increases inside the particle as compared to the free space. This leads to enhanced absorption by internal scatterers.

The dependence of the absorption efficiency factor on the size of particles is given in Fig. 2.5. It follows from this figure, and also from the discussion given above, that the absorption efficiency factor generally increases with the size of particles from its value for Rayleigh scattering ($Q_{abs} = 4Daa/3$) to its asymptotic value equal to $1 - r$ (see Table 2.1), valid as $x \to \infty$ and $\chi x \to \infty$. Oscillations seen in Fig. 2.5 around the size parameter 10 are due to interference effects. They damp down at $x = 20$ because in this case the attenuation of the electromagnetic waves on the diameter of a particle cannot be neglected for the range of χ shown in Fig. 2.5. Generally, the absorption increases with the refractive index. However, the opposite is true in the asymptotic regime because $Q_{ext} = 1 - r$ then and the reflectivity r increases with n (see Table 2.1). This leads to the decrease of the portion of energy, which can penetrate into particles and be absorbed

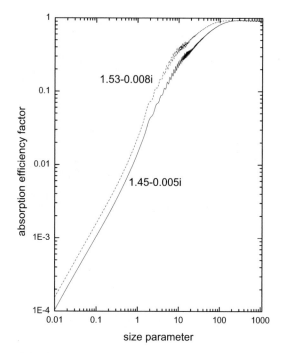

Fig. 2.5. Dependence of the absorption efficiency factor of spherical scatterers on the size parameter $x = ka$ at refractive indices 1.45–0.005i and 1.53–0.008i.

there. Hence, the absorption efficiency factor decreases with n for large, strongly absorbing particles.

2.4 Scattering

The theoretical description of light scattering by atmospheric aerosol is much more complex as compared to absorption and extinction. This is related to the fact that it is not enough just to have information on the aerosol scattering coefficient $k_{sca} = N\overline{C}_{sca}$, which is close to k_{ext} for atmospheric aerosol in most cases; it also is of importance to understand the angular distribution of scattered energy for a given local volume of an aerosol medium.

The dependence of scattering efficiency factor Q_{sca} and also efficiency factors Q_{abs}, Q_{ext} on the size parameter is shown in Fig. 2.6 at $n = 1.45 - 0.005i$. For small size parameters, the absorption is small and $Q_{sca} \approx Q_{ext}$. However, with the growth of particles, Q_{abs} increases and Q_{sca} deviates from Q_{ext} more and more. The scattering efficiency factor reaches its asymptotic value $Q_{sca} = 1 + r$ from above as $x \to \infty, \chi x \to \infty$. Oscillations on curves are due to interference of electromagnetic waves (e.g., diffracted and transmitted). Oscillations damp with the increase of absorption. These oscillations are difficult to observe for natural aerosols because the interference is destroyed by the polydispersity of particles and also because solar light is far a way from an idealized coherent monochromatic incident beam assumed in calculations shown in Fig. 2.6. One can conclude from Fig. 2.6 that the asymptotic regime ($Q_{abs} = 1 - r$) is reached more quickly for the absorp-

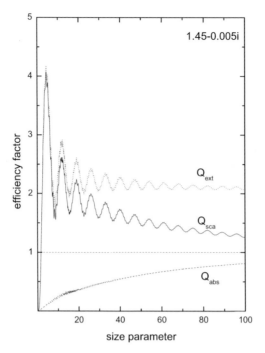

Fig. 2.6. Dependence of Mie efficiency factors on the size parameter $x = ka$ at the refractive index 1.45–0.005i.

tion efficiency factor as compared to the scattering efficiency factor. Clearly, it follows that $Q_{ext} = Q_{abs} + Q_{sca} \rightarrow 2$ from above as $x \rightarrow \infty$. Mie calculations show that Q_{ext} can be described by the following approximate equation in the vicinity of the asymptotic regime: $Q_{ext} = 2(1 + x^{-2/3})$. Accurate estimates of Mie efficiency factors in the vicinity of the asymptotic regime are given by Nussenzveig and Wiscombe (1980) on the basis of the complex angular momentum theory (Nussenzveig, 1992). The color map of Q_{sca} calculated for the same conditions as in Fig. 2.4(a) is given in Fig. 2.7.

In the majority of cases, the distribution of scattered light around the ensemble of particles is azimuthally symmetrical (with respect to the incident light direction). Therefore, this distribution can be described using the single angle equal to zero in the direction of forward scattering and π in the backward direction. This angle is called the scattering angle θ. The notion of the phase function $p(\theta)$ is introduced to describe the angular distribution of the scattered light energy. The value of $p(\theta)d\Omega/4\pi$ gives a conditional probability of light scattering in the solid angle $d\Omega = \sin\theta\,d\theta\,d\phi$. Therefore, it follows that

$$\frac{1}{4\pi} \int_0^{2\pi} d\phi \int_0^{\pi} p(\theta) \sin\theta\,d\theta = 1$$

or

$$\frac{1}{2} \int_0^{\pi} p(\theta) \sin\theta\,d\theta = 1,$$

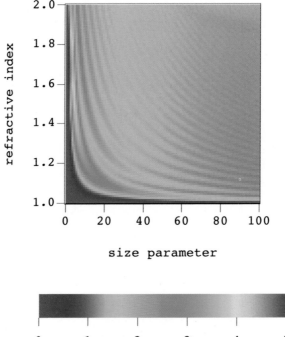

size parameter

scattering efficiency factor

Fig. 2.7. Dependence of the scattering efficiency factor on the size parameter and refractive index of aerosol particles calculated using Mie theory at the imaginary part of the refractive index equal to 0.008.

where we assumed the azimuthal independence of scattering by an aerosol medium. In some aerosol optics applications, the angular distribution of scattered light is represented by just one parameter such as the asymmetry parameter

$$g = \frac{1}{2} \int_0^\pi p(\theta) \cos \theta \sin \theta \, d\theta,$$

the backscattering fraction

$$b = \frac{1}{2} \int_{\pi/2}^\pi p(\theta) \sin \theta \, d\theta,$$

the forward-scattering fraction $f = 1 - b$, their ratio $\eta = b/f$, the average squared scattering angle:

$$\langle \theta^2 \rangle = \int_0^\pi p(\theta) \, \theta^2 \sin \theta \, d\theta,$$

etc. Clearly, these various characteristics are interrelated because they are determined by the same function $p(\theta)$. In particular, it follows for highly extended in the forward direction phase functions: $g = 1 - \langle \theta^2 \rangle / 4$. Andrews et al. (2006) gave the following parameterization of the asymmetry parameter with respect to b:

$$g = 0.9893 - 7.143889b^3 + 7.464439b^2 - 3.96356b.$$

This expression was derived for the special case of the Henyey–Greenstein (HG) phase function given as

$$p(\theta) = \sum_{j=0}^{\infty} (2j + 1)g^j P_j(\cos \theta)$$

or in the closed form:

$$p(\theta) = \frac{1 - g^2}{(1 + g^2 - 2g \cos \theta)^{3/2}}.$$

Here $P_j(\cos \theta)$ is the Legendre polynomial. The values of b and g are usually in the range 0.08–0.18 and 0.5–0.7 (Fiebig and Ogren, 2006), respectively, with most probable values around 0.12 for b and 0.6 for g in the case of atmospheric aerosol.

The HG phase function or linear combinations of these functions with different values of the asymmetry parameter g are used frequently for studies of radiative propagation in the aerosol media. It is difficult to calculate the phase function using the electromagnetic theory for a particle of an arbitrary shape and, even if it can be done for a single particle, the problem of corresponding averaging with respect to the shapes, orientations, and sizes of particles in the local volume remains. This prompts the use of the simplified Heney–Greenstein phase functions in corresponding mathematical modeling of radiative transfer in aerosol media with nonspherical particles.

For spherical particles, the exact solution of the electromagnetic scattering problem is readily available. This enables the calculation of the phase function as the function of the size distribution, the refractive index, and the wavelength of the incident light. In particular, it follows for the case of unpolarized (e.g., solar) incident light that

$$p(\theta) = \frac{2\pi \left(\bar{i}_1(\theta) + \bar{i}_2(\theta) \right)}{k^2 \overline{C}_{sca}},$$

where

$$\bar{i}_s = \int_0^{\infty} i_s(a) f(a) \, da, \quad s = 1, 2, \overline{C}_{sca} = \int_0^{\infty} C_{sca}(a) f(a) \, da$$

and

$$i_1(\theta) = \left| \sum_{l=0}^{\infty} \frac{2l + 1}{l(l + 1)} (a_l \pi_l + b_l \tau_l) \right|^2, \quad i_2(\theta) = \left| \sum_{l=0}^{\infty} \frac{2l + 1}{l(l + 1)} (b_l \pi_l + a_l \tau_l) \right|^2,$$

$$C_{sca} = \frac{2\pi}{k^2} \sum_{l=0}^{\infty} \left(|a_l|^2 + |b_l|^2 \right).$$

The angular functions π_l and τ_l are determined via the associated Legendre function $P_l^1(\cos\theta)$:

$$\pi_l = \frac{P_l^1(\cos\theta)}{\sin\theta} , \tau_l = \frac{dP_l^1(\cos\theta)}{d\theta}.$$

Complex amplitude coefficients a_l and b_l depend on the ratio of the radius of a particle to the wavelength and also on the relative complex refractive index of scatterers as given above. As a matter of fact, expressions given here are valid also for multi-layered spheres except that appropriate amplitude coefficients must then be used. The results presented here make possible the accurate calculation of phase functions of spherical polydispersions.

The expression for the phase function given above can be used for the calculation of the asymmetry parameter. It follows after correspondent analytical integration (van de Hulst, 1957) that

$$g = \frac{4}{x^2 Q_{sca}} \sum_{n=1}^{\infty} \left[\frac{n(n+2)}{n+1} \operatorname{Re}\left(a_n a_{n+1}^* + b_n b_{n+1}^* \right) + \frac{2n+1}{n(n+1)} \operatorname{Re}\left(a_n b_n^* \right) \right].$$

The plots of g calculated using this equation in the same range of n and x as in Fig. 2.1 except at $\chi = 0, \ 0.008$ are given in Fig. 2.8. It follows that generally g increases both with the size of particles and with absorption.

The normalization condition of the phase function is insured by the identity:

$$\overline{C}_{sca} = \frac{\pi}{k^2} \int_0^{\pi} \left(\overline{i}_1 + \overline{i}_2 \right) \sin\theta \, d\theta.$$

The atmospheric aerosol is characterized by the existence of several light scattering modes (e.g., nucleation, fine, and coarse modes). These modes generally have different size distributions and complex spectral reflective indices $m(\lambda)$. Therefore, their phase functions are quite different, as illustrated in Fig. 2.9, derived from the Mie theory at $\lambda = 0.55$ μm using data given in Tables 1.2 and 1.6. Phase functions shown in Fig. 2.9 can be used as building blocks for the construction of different models of atmospheric aerosol. Actually, almost every measured phase function of atmospheric aerosol can be constructed with a sufficient accuracy using various mixtures of functions shown in Fig. 2.9. The following equations can be used to find the optical characteristics of a mixture of three aerosol components:

$$k_{ext} = \left(v_1 k_{ext}^{(1)} + v_2 k_{ext}^{(2)} + v_3 k_{ext}^{(3)} \right) v,$$

where $k_{ext}^{(i)} = \overline{C}_{ext}^i / \overline{V}^{(i)}$ is the volumetric extinction coefficient of a given aerosol mode and $\overline{V}^{(i)} = (4\pi/3) \int_0^{\infty} a^3 f_i(a) \, da$ is the average volume of aerosol particles, $f_i(a)$ is the PSD of the i-th aerosol mode. The values of $v_1, v_2, v_3 (v_1 + v_2 + v_3 = 1)$ give the volume mixing ratios (e.g., $v_1 = 0.29(\text{FM}), v_2 = 0.7(\text{CM}), v_3 = 0.01(\text{SA})$ for continental aerosols, see

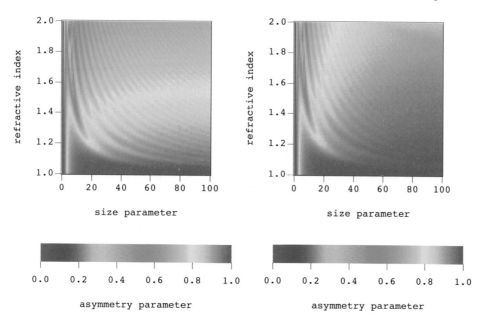

Fig. 2.8. Dependence of the asymmetry parameter on the size parameter and refractive index of aerosol particles calculated using Mie theory at the imaginary part of refractive index equal to 0.0 (a) and 0.008 (b).

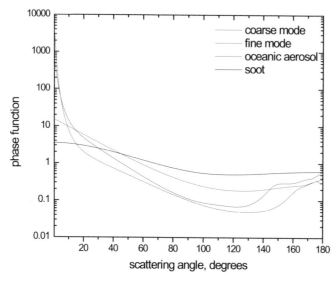

Fig. 2.9. Phase functions of selected aerosol types.

Table 1.7) and v is the total volumetric concentration of particles in the unit aerosol volume. The same mixing rule is applied to the absorption and scattering coefficients. The mixing rule for the phase function of a mixture has the following form:

$$p(\theta) = \frac{v_1 k_{\text{sca}}^{(1)} p^{(1)}(\theta) + v_2 k_{\text{sca}}^{(2)} p^{(2)}(\theta) + v_3 k_{\text{sca}}^{(3)} p^{(3)}(\theta)}{v_1 k_{\text{sca}}^{(1)} + v_2 k_{\text{sca}}^{(2)} + v_3 k_{\text{sca}}^{(3)}},$$

which underlines the fact that the scattered light intensities and not the phase functions must be added in the process of creation of the mixture. In a similar way, we have for the asymmetry parameter:

$$g = \frac{v_1 k_{\text{sca}}^{(1)} g^{(1)} + v_2 k_{\text{sca}}^{(2)} g^{(2)} + v_3 k_{\text{sca}}^{(3)} g^{(3)}}{v_1 k_{\text{sca}}^{(1)} + v_2 k_{\text{sca}}^{(2)} + v_3 k_{\text{sca}}^{(3)}}.$$

The equations given above are easily generalized for any number of aerosol models i. In particular, one obtains for a bimodal model:

$$k_{\text{ext}} = (v_1 k_{\text{ext}}^{(1)} + (1 - v_1) k_{\text{ext}}^{(2)}) v, \quad p(\theta) = \varepsilon p^{(1)}(\theta) + (1 - \varepsilon) p^{(2)}(\theta),$$

where the parameter $\varepsilon \in [0, 1]$ is defined as

$$\varepsilon = \frac{v_1 k_{\text{sca}}^{(1)}}{v_1 k_{\text{sca}}^{(1)} + (1 - v_1) k_{\text{sca}}^{(2)}}.$$

The phase functions of continental and maritime aerosol shown in Fig. 2.10 have been obtained from phase functions presented in Fig. 2.9 using volume mixing ratios as specified in Table 1.7. The integral light scattering characteristics derived from Mie theory for the cases illustrated in Figs. 2.9 and 2.10 are shown in Table 2.2. The results presented in Table 2.2 can be used in theoretical modeling of radiative transfer properties of atmospheric aerosols. However, one must remember that aerosol properties vary and, therefore, they can in reality deviate considerably from the results given in Table 2.2 (see, e.g., Dubovik et al., 2002).

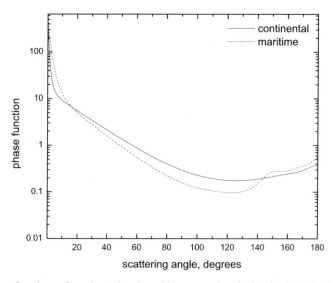

Fig. 2.10. Phase functions of continental and maritime aerosols calculated using Mie theory.

Table 2.2. Microphysical and integral light scattering characteristics of different aerosol types at $\lambda = 550$ nm

Aerosol type	a_0, nm	σ	n	χ	b	g	ω_0
Fine	5	1.095 27	1.53	0.006	0.1098	0.6281	0.9569
Coarse	500	1.095 27	1.53	0.008	0.0380	0.8689	0.6659
Oceanic	300	0.920 28	1.38	0.0	0.0642	0.7851	1.0
Soot	11.8	0.79	1.75	0.44	0.2681	0.3366	0.2088
Continental	–	–	–	–	0.1057	0.6451	0.8816
Maritime	–	–	–	–	0.0753	0.7430	0.9890

Note that the integration with respect to the particle size distribution for the data shown in Figs. 2.9, 2.10 and Table 2.2 was performed in the range of radii 0.005–20 μm. Mathematically speaking, the upper radius must be increased considerably for coarse and oceanic aerosols. However, it is believed that the cutoff at large radii is really necessary for a better representation of realistic microphysical and optical characteristics of atmospheric aerosols. This also means that it is not enough to give the pair (a_0, σ) in the output of aerosol particle size distribution retrieval algorithms. The upper and lower integration limits also must be specified. This will enable the derivation of the average radius $\bar{a}$ and the standard deviation Δ for the retrieved PSD. These characteristics can be used for the intercomparison of retrievals based on different assumptions on the PSD type (Junge, gamma PSD, etc.). Actually, it is advised that the pair $(\bar{a}, \Delta)$ is reported in the outputs of corresponding aerosol inversion algorithms. This is not the case at the moment, although these characteristics represent the statistical properties of an ensemble of particles under study in the most direct way and can be used for the intercomparison of aerosol properties derived by different groups using diverse retrieval and measurement techniques.

2.5 Polarization

Light is composed of electromagnetic waves oscillating with a high frequency (typically, 10^{15} oscillations per second). An important property of any electromagnetic wave is its polarization. Unpolarized light can be transformed to almost 100 % polarized light using the phenomena of light reflection, transmission, and scattering. A typical example is the reflection from a plane surface at the Brewster angle equal to $\arctan(n)$. Here n is the relative refractive index of a surface. The light reflected at this angle becomes completely polarized in the plane perpendicular to the plane of incidence holding the incident and reflected beams. Dipole scattering produces 100 % linearly polarized at the scattering angle equal to $\pi/2$. Clearly, polarization effects play an important role in atmospheric optics in general and in aerosol optics in particular (Gorchakov, 1966). They are sensitive to both the size and refractive index of particles. Also polarization characteristics of reflected light are influenced by the shape and internal structure of scatterers. Therefore, they enable the solution of a number of inverse aerosol optics problems including aerosol remote sensing from aircraft, space and ground (Chowdhary et al., 2005).

Usually the polarization of a scattered light beam is represented in terms of Stokes parameters defined as

$$I = I_1 + I_r, Q = I_1 - I_r, U = J_1 + J_r, V = i(J_1 - J_r),$$

where $I_1 = \langle E_1 E_1^* \rangle, I_r = \langle E_r E_r^* \rangle, J_1 = \langle E_1 E_r^* \rangle, J_r = \langle E_r E_1^* \rangle$. Here $E_{1(r)}$ are components of the electric vector parallel (1) and perpendicular (r) to the scattering plane and the averaging with respect to the spatial and temporal constant of a receiver is denoted by the angle brackets. The common constant multiplier in the definition of Stokes parameters is omitted. The Stokes parameters enable the determination of the degree of linear polarization:

$$P_1 = \frac{\sqrt{Q^2 + U^2}}{I}.$$

It follows at $U = 0$: $P_1 = -Q/I$. The choice of the sign is such that the degree of polarization for molecular scattering ($Q < 0$) is the positive number. Also one can define the total degree of polarization:

$$P = \frac{\sqrt{Q^2 + U^2 + V^2}}{I}$$

and the degree of circular polarization $P_c = V/I$. Clearly, it follows that

$$P = \sqrt{P_c^2 + P_1^2}.$$

The Stokes vector of the partially polarized light beam can be presented as the sum of the Stokes vector $S_p(PI, Q, U, V)$ of completely polarized light and the Stokes vector $S_u((1 - P)I, 0, 0, 0)$ of unpolarized light. The ellipticity of radiation is defined as $\beta = (1/2) \arcsin(V/I)$ and $\chi = (1/2) \arctan(U/Q)$ is the angle describing the orientation of the ellipse of polarization. The pair (β, χ) is used in the optical diagnostic of various surfaces (e.g., the determination of their optical constants). In a similar way they can be used in the development of the aerosol ellipsometry either in the laboratory or in field studies (Kokhanovsky, 2003). The technique is highly sensitive both to real and imaginary parts of the refractive index of scatterers.

Analytical calculations in the field of polarization optics are simplified, if not the Stokes vector but rather the density matrix

$$\rho = \vec{E} \otimes \vec{E}^+$$

is used. The plus sign means the simultaneous operation of transportation and conjugation. The direct product as given above means that ρ is a 2×2 matrix of the following form:

$$\rho = \begin{pmatrix} I_1 & J_1 \\ J_r & I_r \end{pmatrix}.$$

Now we take into account that electric vectors of scattered (s) and incident (i) waves can be represented as follows:

$$\begin{pmatrix} E_1^s \\ E_r^s \end{pmatrix} = A\hat{s} \begin{pmatrix} E_1^i \\ E_r^i \end{pmatrix},$$

where $A = -i \exp(ik(z - r))/kr$, r is the distance to the observation point and z is the coordinate along the direction of light incidence. $\hat{s}$ is the so-called amplitude scattering matrix, which has the following form for the case studied (van de Hulst, 1957):

$$\hat{s} = \begin{pmatrix} s_{11} & 0 \\ 0 & s_{22} \end{pmatrix},$$

where

$$s_{11} = \sum_{l=0}^{\infty} \frac{2l+1}{l(l+1)} (b_1 \pi_l + a_l \tau_l), \quad s_{22} = \sum_{l=0}^{\infty} \frac{2l+1}{l(l+1)} (a_l \pi_l + b_l \tau_l).$$

Taking into account the definition of the density matrix, we derive:

$$\rho \equiv \vec{E} \otimes \vec{E}^+ = \frac{\hat{s}\vec{E}_i \otimes \left(\hat{s}\vec{E}_i\right)^+}{k^2 r^2}$$

or

$$\rho = \frac{\hat{s}\rho_i \hat{s}^+}{k^2 r^2}.$$

This expression shows how the density matrix of the scattered light beam ρ can be represented in terms of the density matrix of the incident light beam ρ_i. Let us find a so-called scattering matrix $\hat{M}$, which transforms the Stokes vector of incident light $\vec{S}_0$ to the Stokes vector of the scattered light $\vec{S}$:

$$\vec{S} = \frac{\hat{F}}{k^2 r^2} \vec{S}_0.$$

For this, we note that the density matrix can be represented in the following form:

$$\rho = \frac{1}{2} [I\sigma_1 + Q\sigma_2 + U\sigma_3 + V\sigma_4],$$

where

$$\hat{\sigma}_1 = \begin{pmatrix} 1 & 0 \\ 0 & 1 \end{pmatrix}, \hat{\sigma}_2 = \begin{pmatrix} 1 & 0 \\ 0 & -1 \end{pmatrix}, \hat{\sigma}_3 = \begin{pmatrix} 0 & 1 \\ 1 & 0 \end{pmatrix}, \hat{\sigma}_4 = \begin{pmatrix} 0 & -i \\ i & 0 \end{pmatrix}.$$

Therefore, we have for the components of the Stokes vector ($S_1 = I$, $S_2 = Q$, $S_3 = U$, $S_4 = V$):

$$S_j = \text{Tr}(\hat{\sigma}_j \hat{\rho}),$$

where Tr is the trace operation. Also it follows that

$$\rho_0 = \frac{1}{2} \sum_{k=1}^{4} I_{0k} \hat{\sigma}_k$$

and, therefore,

$$S_j = \frac{1}{k^2 r^2} F_{jk} S_{0k},$$

where

$$F_{jk} = \frac{1}{2} \text{Tr}\left(\hat{\sigma}_j \hat{s} \hat{\sigma}_k \hat{s}^+\right)$$

are elements of $4 * 4$ scattering matrix $\hat{F}$. This establishes the law of transformation of the Stokes vector of the incident light due to the scattering process. We see that the dimensionless 4×4 transformation matrix $\hat{F}$ is determined solely by the $2 * 2$ amplitude scattering matrix $\hat{s}$. Simple calculations give for nonzero elements of this matrix for spheres:

$$F_{11} = F_{22} = \frac{1}{2}(i_1 + i_2),$$

$$F_{12} = F_{21} = \frac{1}{2}(i_1 - i_2),$$

$$F_{33} = F_{44} = \text{Re}\left(s_{11}s_{22}^*\right),$$

$$F_{34} = -F_{43} = \text{Im}\left(s_{11}s_{22}^*\right),$$

where $i_1 = s_{11}s_{11}^*$, $i_2 = s_{22}s_{22}^*$. One can represent s_{nn} in the exponential form: $s_{nn} = \varsigma_n \exp(-i\varphi_n)$, where ς_n and φ_n are real numbers and $n = 1, 2$. Then it follows: $F_{11} = (\varsigma_1^2 + \varsigma_2^2)/2$, $F_{12} = (\varsigma_1^2 - \varsigma_2^2)/2$, $F_{33} = \varsigma_1\varsigma_2 \cos\phi$, $F_{34} = \varsigma_1\varsigma_2 \sin\phi$, where $\phi = \varphi_1 - \varphi_2$. It can be easily verified that $F_{11}^2 = F_{12}^2 + F_{34}^2 + F_{44}^2$, which reflects the important property of monodispersed spherical particles: they do not change the total degree of polarization or, therefore, the entropy of incident completely polarized light beam. This is not the case for nonspherical particles or for spherical polydispersions. Note that it follows for randomly oriented nonspherical particles that $F_{22} \neq F_{11}$, $F_{33} \neq F_{44}$.

In multiple light scattering studies, not the matrix $\hat{F}$ but the normalized phase matrix

$$\hat{P} = \frac{4\pi}{k^2 C_{\text{sca}}} \hat{F}$$

is usually used. The element P_{11} coincides with the phase function. Also it is useful to introduce the normalized scattering matrix having elements $f_{ij} = F_{ij}/F_{11}$. It follows for spheres that

$$f_{11} = f_{22} = 1,$$

$$f_{12} = f_{21} = \frac{i_1 - i_2}{i_1 + i_2},$$

$$f_{33} = f_{44} = \frac{\text{Re}\left(S_1 S_2^*\right)}{i_1 + i_2},$$

$$f_{34} = -f_{43} = \frac{\text{Im}\left(S_1 S_2^*\right)}{i_1 + i_2}$$

with all other elements equal to zero. It follows for the monodispersed spherical particles from the discussion given above that $f_{12}^2 + f_{34}^2 + f_{44}^2 = 1$. This means that for a complete description of light scattering process by a single spherical aerosol particle one needs to have just three functions, e.g., $p(\theta)$, $f_{12}(\theta)$, and $f_{44}(\theta)$. Then it follows that $f_{34}(\theta) = \pm\sqrt{1 - f_{12}^2 - f_{44}^2}$ and the sign coincides with the sign of ϕ. So the measurements of angular distributions of light scattered intensity and degree of polarization for unpolarized incident light (f_{12}) and also the degree of circular polarization (f_{44}) of incident circularly polarized light (Kokhanovsky, 2003) are needed. The underlying reason for this is quite clear. Indeed, the scattering process is described by the complex amplitude scattering matrix elements $s_{11}(\theta)$ and $s_{22}(\theta)$ in the case of spherical particles composed of isotropic substances (e.g., characterized by the scalar and not the tensor dielectric permittivity). So, in total four numbers are needed, for example, in microwave scattering experiments. However, the nature of optical measurements is such that Stokes vector components and not the amplitude scattering matrix elements can be measured (Stokes, 1852; Perrin, 1942, Chandrasekhar, 1950; Rozenberg, 1955). These elements are quadratic with respect to the electromagnetic field. Therefore, only modules of $s_{11}(\theta)$ and $s_{22}(\theta)$ and their relative phase as discussed above enter the theory of scattered light fields.

We present the values of the phase function $P_{11}(\theta)$, the degree of polarization $P_1 = -f_{12}$, and also $f_{33} = f_{44}, f_{34} = -f_{43}$ in Fig. 2.11. Calculations have been performed at $n = 1.53 - 0.008i$ for various sizes of spherical monodispersed particles ($x \in [0.01, 100]$). One concludes that values of the phase function are quite small in the range of scattering angles 90–150 degrees, with some increase and also larger variability depending on the size in the backscattering region. The white color in Fig. 2.11(a) corresponds to values of the phase function larger than one. This is mostly the case for forward-scattering region. The degree of polarization is quite low (green color in Fig. 2.11(b)) except in the vicinity of the backscattering direction, where P_1 switches the sign depending on x, and also at the rainbow region (around $\theta \sim 100°$), where the area of high positive polarization is located. One concludes from Fig. 2.11(c) that the degree of polarization of incident circularly polarized light is hardly changed in the forward direction. It changes sign (opposite sense of rotation as compared to the incident beam) around the 125-degree scattering angle (blue color in Fig. 2.11(c)). The element f_{34} shown in Fig. 2.11(d) gives the linear-to-circular light polarization conversion strength (Kokhanovsky, 2003) by aerosol particles. It follows from this figure that the probability of such conversion is low (except the backscattering region, see blue color).

The elements of the normalized scattering matrix of water aerosol calculated using Mie theory for spherical polydispersions and also measured in the laboratory at the laser wavelengths 441.6 and 632.8 nm are given in Fig. 2.12. It follows from this figure that Mie theory can be successfully used for the theoretical interpretation of corresponding polarimetric experiments in the field of aerosol optics. Lines correspond to calculations for the lognormal PSD with an effective radius of 1.1 μm and an effective variance equal to 0.5 at a refractive index equal to 1.33.

In situ measurements of the aerosol angular scattering coefficient $D(km^{-1}sr^{-1})$ and also the normalized phase matrix elements $\tilde{f}_{21} = f_{12}$ $\tilde{f}_{33} = f_{33}$ $\tilde{f}_{43} = -f_{34}$ are given in Fig. 2.13. There are large differences in the angular functions given in Figs. 2.12 and 2.13, which are due to different sizes and refractive indices of particles. Note that the measurements presented at the left part of Fig. 2.13 are best fitted by the Deirmendjian's haze L model of PSD (Deirmendjian, 1969) except with the change of mode radius to 0.11 μm and a refractive index of 1.37. Measurements at the right part of Fig. 2.13 are

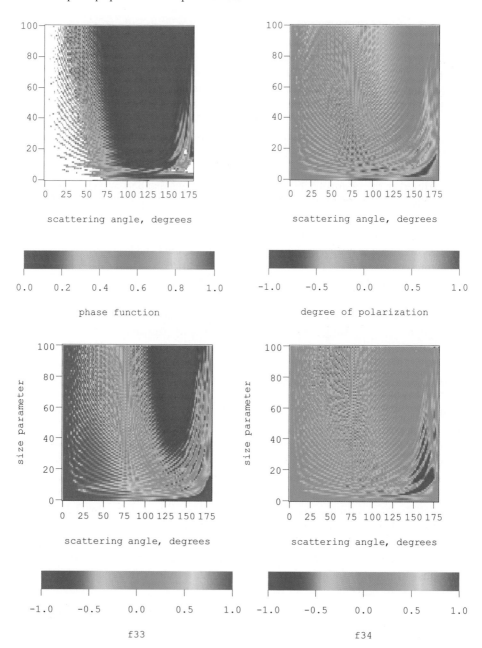

Fig. 2.11. Phase function (a), degree of polarization (b), f_{33} (c), f_{34} (d) for the refractive index 1.53–0.008i and $x \in [0.01,100]$.

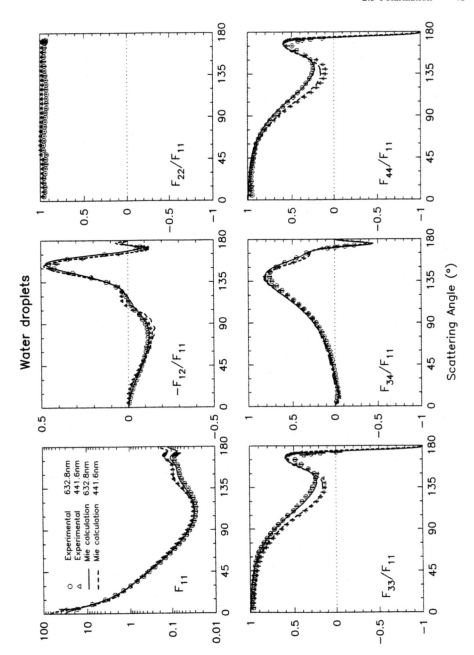

Fig. 2.12. Experimentally measured phase functions and normalized phase matrix elements of water aerosol (symbols) at laser wavelengths 441.6 and 632.8 nm. The lines give the results of the fit using Mie theory (Volten et al., 2001).

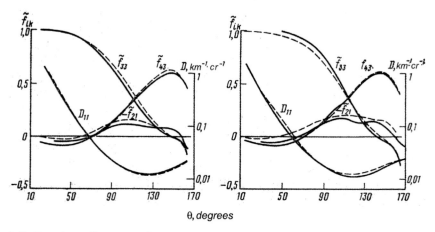

θ, degrees

Fig. 2.13. Experimentally measured phase function and normalized phase matrix elements of atmospheric haze (solid lines) at 550 nm. The broken lines give the results of the fit using Mie theory (Gorchakov et al., 1976).

best fitted using slightly smaller values of a_m and n (0.1 μm and 1.35, respectively). The good fit of the angular functions $D(\theta)$ can be achieved assuming the scattering coefficient $k_{sca} = 1.5$ km^{-1} for the measurements at the left part of the figure and $k_{sca} = 1.7$ km^{-1} at the right part of the figure. One can see that the fit of measurements by Mie theory is better in Fig. 2.12 as compared to Fig. 2.13. This can be explained by the fact that monomodal spherical polydispersions are not adequate for the explanation of the experimental measurements shown in Fig. 2.13. The use of bimodal distributions of polydispersions having different mode radii and also refractive indices can improve the fit. The presence of non-spherical particles cannot be ruled out either. The measurements shown in Fig. 2.13 have

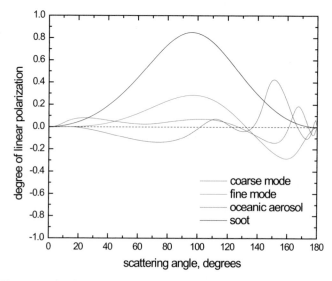

Fig. 2.14(a). The same as in Fig. 2.9 except for the degree of linear polarization of singly scattered light assuming unpolarized light illumination conditions.

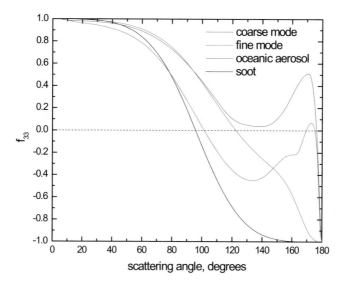

Fig. 2.14(b). The same as in Fig. 2.9 except for f_{33}.

been performed *in situ* in a rural area close to Moscow under conditions of a stable atmospheric haze (Gorchakov et al., 1976).

The degree of polarization $P \equiv -f_{12}$ for the aerosol models shown in Table 2.2 is given in Fig. 2.14(a). It follows that the value of P is larger for fine mode aerosols and also for soot, which shows a polarization curve similar to that for molecular scattering, which is given by $\sin^2\theta/(1 + \cos^2\theta)$. The polarization curves given in Fig. 2.14(a) are very different and, therefore, they can be used for the identification of the predominant aerosol type. Elements f_{33} and f_{34} are shown in Figs. 2.14(b) and (c), respectively. Calculations were

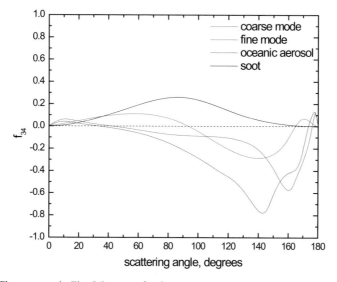

Fig. 2.14(c). The same as in Fig. 2.9 except for f_{34}.

carried out for the same conditions as in Fig. 2.14(a). The physical meaning of the element $f_{44} = f_{33}$ is the degree of circular polarization P_c of scattered light under the illumination of an aerosol medium by right-hand completely circular polarized light (Kokhanovsky, 2003). It follows from Fig. 2.14(b) that the scattering process in the forward direction hardly changes the degree of polarization or the direction of a rotation of circularly polarized light beam. At exactly 180 degrees, the absolute value of the degree of circular polarization is equal to one, as it is for incident light. However, the sense of rotation is opposite to that in the incident beam. The value of P_c vanishes at some scattering angles depending on the aerosol model (see Fig. 2.14(b)). This feature can be used for the identification of the aerosol type. In particular, the value of P_c for oceanic aerosol undergoes rapid changes at scattering angles larger than 160 degrees.

The value of f_{34} (see Fig. 2.14(c)) can be also interpreted as the degree of circular polarization of scattered light for incident linearly polarized light (with the azimuth -45 degrees). Therefore, f_{34} shows the linear-to-circular polarized light conversion efficiency. Usually, such a conversion has a low efficiency with $|P_c|$ smaller than 30 % in the forward hemisphere (see Fig. 2.14(c)). However, in the backward hemisphere, there are regions, where the value of $|P_c|$ is quite large and can reach 0.8 at the scattering angle 142 degrees, for example, in the vicinity of the angular range, where the rainbow occurs. Maxima of $|P_c|$ shown in Fig. 2.14(c) for oceanic and coarse aerosol modes are shifted with respect to each other due to the different values of the real part of the refractive index of soil and oceanic aerosols.

We show the degree of polarization given by $-f_{12}$ and also f_{33}, f_{34} for continental and maritime aerosols calculated using the data shown in Figs. 2.14(a), (b) and (c) and mixing ratios presented in Table 2.2 and Figs. 2.15(a), (b) and (c). It follows from comparisons of Figs. 2.15 and Figs. 2.14 that the scattering matrix of continental aerosol behaves similar by to the fine mode aerosol scattering matrix and that that for the maritime aerosol is very

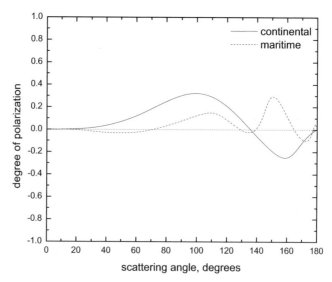

Fig. 2.15(a). The same as in Fig. 2.10 except for the degree of linear polarization of singly scattered light assuming unpolarized light illumination conditions.

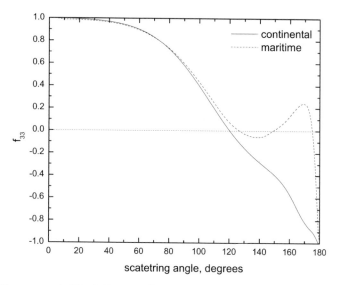

Fig. 2.15(b). The same as in Fig. 2.10 except for f_{33}.

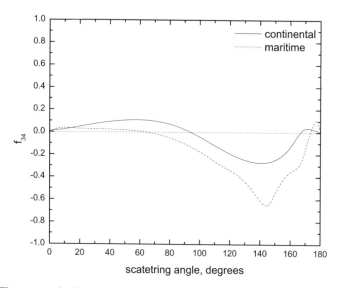

Fig. 2.15(c). The same as in Fig. 2.10 except for f_{34}.

close to the oceanic aerosol model. Also it follows that the behavior of scattering matrix elements differs for continental and maritime aerosol considerably (especially, in the backward hemisphere). This can be used, for example, for the identification of the aerosol type.

It follows from Fig. 2.15(b) that the degree of circular polarization vanishes, for example, at 175 degrees for maritime aerosol. This, however, does not mean that the light becomes completely unpolarized. Apart from the depolarization, a process of circular-to-linear polarization conversion takes place at this angle.

Chapter 3. Multiple light scattering in aerosol media

3.1 Radiative transfer equation

The average sizes of most aerosol particles are of the order of the visible light wavelength. This means that optical methods are very suitable for studies of atmospheric aerosol. However, apart from the nonsphericity and also inhomogeneity of particles yet another problem arises. One needs to account for multiple light scattering to characterize processes of light transmission, reflection and diffusion in aerosol layers. This problem is quite complex in mathematical terms (Mishchenko et al., 2002). It is treated usually in the simplified framework of radiative transfer theory. It means that instead of manipulating with electromagnetic fields $\vec{E}$, one considers the transformation of the Stokes vector of the incident light by an aerosol medium. This enables the description of almost all possible experimental measurements in the field of aerosol optics. We start from the consideration of the scalar radiative transfer equation for the intensity of light field ignoring polarization effects. Also it is assumed that the scattering medium is isotropic, that scatterers are situated at large distances one from another and that there are no nonlinear effects (e.g. dependencies $k_{ext}(I_t)$, I_t is the light intensity), time-dependent effects (e.g., propagation of laser pulses) and frequency change in the scattering processes. Also the possible effects of stimulated emission (e.g., lasing in aerosol media) are omitted. Even with so many simplified assumptions, the radiative transfer equation (RTE) has the following complicated form:

$$(\vec{n}\,\overrightarrow{grad})I_t(\vec{r},\vec{n}) = -k_{ext}I_t(\vec{r},\vec{n}) + \frac{k_{sca}}{4\pi}\int_{4\pi} p(\vec{n},\vec{n}')I_t(\vec{r},\vec{n}')\,d\vec{n}' + B_0(\vec{r},\vec{n}),$$

where $\vec{r} = x\vec{l}_x + y\vec{l}_y + z\vec{l}_z$ is the radius-vector of the observation point, the vector $\vec{n} = l\vec{e}_x + m\vec{e}_y + n\vec{e}_z$ determines the direction of beam with the intensity I_t, $B_0(\vec{r},\vec{n})$ is the internal source function. This function describes internal (e.g., thermal) sources of radiation. The physical meaning of this equation is quite simple: the change in the light intensity (the derivative) in the direction $\vec{n}$ at the point with the radius-vector $\vec{r}$ is due to the extinction $(-k_{ext}I_t)$ and scattering (the integral)/emission (B_0) processes. The phase function $p(\vec{n},\vec{n}')$ describes the strength of light scattering from the direction $\vec{n}'$ to the direction $\vec{n}$ by a local volume of an aerosol medium and the corresponding integral accounts for the total contribution of the scattered light from all directions to the direction $\vec{n}$ at the point with the radius-vector $\vec{r}$.

The physical nature of propagating particles (e.g., neutrons, electrons, photons, etc.) in various media can be different and, therefore, the physical meaning of k_{sca}, k_{ext}, and $p(\theta)$ and also physical theories used for their calculations can be quite different, although the multiple light scattering effects are correctly described by the RTE given above independently of the physical nature of the problem.

This integro-differential equation, which involves integration with respect to the vector $\vec{n}'$ and the directional derivative along the path L

$$(\vec{n} \cdot \vec{grad}) I_t(\vec{r}, \vec{n}) \equiv \frac{dI_t(\vec{r}, \vec{n})}{dL}$$

enables the description of 3-D radiative transfer in aerosol media of arbitrary shapes. This equation can be solved using either Monte Carlo techniques or various grid techniques, which substitute summation for the integration with the reduction of the general problem to the solution of the system of inhomogeneous differential equations. There are a number of standard mathematical techniques and codes available online for the solution of differential equations and their systems (e.g., the 3-D radiative transfer code of Evans (1998) is available online; see, for example, http://en.wikipedia.org/wiki/List_of_atmospheric_radiative_transfer_codes).

To better understand the internal structure of radiative transfer theory, we consider the much simpler case of 1-D radiative transfer. This problem not only is of importance from the methodological point of view but also has a lot of practical applications. In particular, current satellite aerosol remote sensing from space is based exclusively on 1-D RTE. In a way, 1-D theory serves a role similar to that of Mie theory in single light scattering by particles. It enables fast calculations and captures the main physical dependencies. The analytical theory and numerical procedures of 1-D radiative transfer are highly developed and mature enough to use in the solution of most practical problems arising in field of optical engineering and also in the related areas of applied optics.

1-D theory assumes that a horizontally homogeneous plane-parallel layer of a finite optical thickness τ_0 can be substituted for the scattering aerosol layer. The value of τ_0 is defined as an integral of the extinction coefficient along the vertical coordinate inside the scattering layer. The scattering layer is infinite in the horizontal direction. Its microphysical and optical properties can vary in the vertical direction, however. Here, we will consider only a simplified case of a vertically homogeneous layer, which can be characterized by a single value of the single scattering albedo, the extinction coefficient k_{ext} and the phase function $p(\theta)$, which is the same at any point in the medium. Therefore, it follows that $\tau_0 = k_{ext} L$, where L is the geometrical thickness of the layer. Complications arising due to vertical inhomogeneity of the medium are addressed in a comprehensive work of Yanovitskij (1997). Also it is assumed that a layer is uniformly illuminated on its top by a unidirectional light beam. So the case of illumination by narrow beams (e.g., laser) at a given point of an aerosol boundary is not considered. For simplicity, we assume that there are no thermal sources of radiation in the medium.

Then the main equation of the theory can be written in the following form (Sobolev, 1975):

$$\cos \vartheta \frac{dI_t(\tau, \vartheta, \phi)}{d\tau} = -I_t(\tau, \vartheta, \phi) + B_t(\tau, \vartheta, \phi),$$

where

$$B_t(\tau, \vartheta, \phi) = \frac{\omega_0}{4\pi} \int_0^{2\pi} d\phi' \int_0^{\pi} I_t(\tau, \vartheta', \phi') p(\theta') \sin \vartheta' \, d\vartheta'$$

is the source function and $I_t(\tau, \vartheta, \phi)$ is the total light intensity at the optical thickness τ in the direction (ϑ, ϕ). The complication arises due to the fact that B_t depends on the light intensity I_t. Here $\omega_0 = k_{sca}/k_{ext}$ is the single scattering albedo, $p(\theta)$ is the phase function, $\tau = k_{ext}z$ is the optical depth, z is the vertical coordinate (see Fig. 3.1 for the definition of the coordinate system), ϑ is the observation angle, and ϕ is the azimuth. The phase function $p(\theta)$ describes the conditional probability of light scattering from the direction specified by the pair (ϑ', ϕ') to the direction (ϑ, ϕ). One can derive, using spherical trigonometry:

$$\cos \theta' = \cos \vartheta \cos \vartheta' + \sin \vartheta \sin \vartheta' \cos(\phi - \phi').$$

Therefore, the physical problem of light diffusion in an aerosol medium is reduced to the solution of the integro-differential equation given above for *a priori* known values of ω_0 and the phase function $p(\theta)$. The solution must be found in an arbitrary direction specified by the pair (ϑ, ϕ) at any value of τ inside of the medium and also at its boundaries (e.g., $\tau = 0$ at the illuminated top of an aerosol layer and $\tau = \tau_0$ at the bottom of the layer; see Fig. 3.1). The size, shape, internal structure, and refractive index of aerosol particles determine the values of $\tau_0 = k_{ext}L$, ω_0 and the phase function $p(\theta)$. These parameters can be calculated using, for example, Mie theory, as described in the previous chapter, both for monodispersed spheres and also for the spherical polydispersions including multimodal size distributions of aerosols having different physical origins and chemical compositions. Also advanced mathematical theories for cases of nonspherical particles can be used (Mishchenko et al., 2002). This separation of calculations of single scattering and absorption

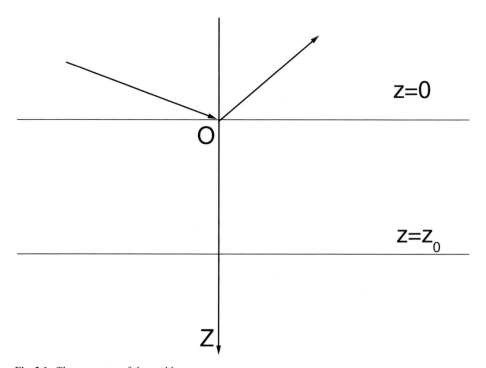

Fig. 3.1. The geometry of the problem.

effects for the local unit volume of an aerosol layer from the calculation of global optical properties such as the total light scattering intensity and, therefore, also transmission and reflection coefficients of an aerosol layer is an important feature of the radiative transfer theory rooted in a number of simplified assumptions required in the derivation of RTE from the Maxwell theory (Mishchenko, 2002). One such assumption is the possibility of neglecting close-packed media effects (Kokhanovsky, 2004a), which is fortunately the case for aerosol optics problems. For instance, it follows that the volume concentration of aerosol particles, defined as the product of their number concentration and the average volume, is just $\approx 4 \cdot 10^{-9}$, if one assumes $N = 10^6$ particles of a radius 0.1 μm in a cubic centimeter of air. Usually the values of N and, therefore, their volumetric concentration are much smaller than that. So in reality, the volumetric concentration of particles and, therefore, the distance between them are much larger than the diameter of an aerosol particle and also the wavelength of visible light. These important conditions allow for a considerable simplification of aerosol optics problems as compared to those arising, for example, in the problems of radiative transfer in close-packed media such as soil, snow, and whitecaps. As a matter of fact, one can perform studies in the field of radiative transfer not referring to the exact physical meaning of local characteristics such as ω_0, τ_0, and $p(\theta)$. For instance, as emphasized above the same equation can be applied for studies of the neutron transport and also for understanding the diffusion of fast charged particles in condensed matter. The only difference will be that constants ω_0 and τ_0 and also the function $p(\theta)$ are determined by physical laws different from those underlying light scattering by aerosol particles. This multidisciplinary nature of the radiative transfer theory and its applicability to a great range of physical problems is the most important underlying reason for great developments in the field achieved in the 20th century with contributions from scientists working in very diverse research fields.

3.2 The diffuse light intensity

It is of advantage to represent the total light intensity in the following form:

$$I_t(\tau, \vartheta, \phi) = \pi I_0 \, \exp\left(-\frac{\tau}{\cos \vartheta_0}\right) \delta(\vec{\Omega} - \vec{\Omega}_0) + I(\tau, \vartheta, \phi)$$

where πI_0 is the incident radiative flux per unit area normal to the beam. The first term represents the attenuation of the incident direct light beam in the direction $\vec{\Omega}(\vartheta, \phi)$ coinciding with the direction $\vec{\Omega}_0(\vartheta, \phi)$ of an incident beam, which is ensured by the delta function $\delta(\vec{\Omega} - \vec{\Omega}_0)$. The second term represents the so-called diffuse light intensity $I(\tau, \vartheta, \phi)$. Therefore, the equation given above enables the separation of the total intensity into two components: the diffuse light intensity, which is a slow changing function of angles, and a very peaked angular function describing the direct attenuated light beam. Such a separation enables the simplification of corresponding numerical procedures for the calculation of light fields in aerosol media as compared to the direct calculation of the total intensity $I_t(\tau, \vartheta, \phi)$.

It is easy to derive the following equation for the diffuse light intensity from the general RTE using the separation procedure underlined above and also the well known properties of the delta function:

$$\cos\vartheta \frac{dI(\tau,\vartheta,\phi)}{d\tau} = -I(\tau,\vartheta,\phi) + B(\tau,\vartheta,\phi),$$

where

$$B(\tau,\vartheta,\phi) = \frac{\omega_0}{4\pi} \int_0^{2\pi} d\phi' \int_0^{\pi} I(\tau,\vartheta',\phi')p(\theta) \sin\vartheta' \, d\vartheta' + \frac{\omega_0 I_0}{4} p(\theta) \exp(-\tau/\cos\vartheta_0),$$

where $\cos\theta = \cos\vartheta\cos\vartheta_0 + \sin\vartheta\sin\vartheta_0\cos(\phi)$ and we assumed that $\phi_0 = 0$. This is the main equation to be solved. For the completeness of the mathematical problem, we need to specify the boundary conditions. We will assume that the aerosol layer is situated in vacuum over a black underlying surface. This means that the diffuse light has no chance to enter the top of the layer ($\tau = 0$) from above and the bottom of the layer ($\tau = \tau_0$) from below. Therefore, boundary conditions can be presented in the following form:

$$I(0,\vartheta,\vartheta_0,\varphi) = 0 \text{ at } \vartheta < \frac{\pi}{2},$$

$$I(\tau_0,\vartheta,\vartheta_0,\varphi) = 0 \text{ at } \vartheta > \frac{\pi}{2}.$$

The differential equation for the diffuse intensity given above can be solved analytically. The answer is:

$$I(\tau,\eta,\xi,\phi) = \frac{e^{-\tau/\eta}}{\eta} \int_0^{\tau} B(\tau',\eta,\xi,\phi) e^{\tau'/\eta} \, d\tau' \text{ at } \eta > 0$$

for the downward intensity and

$$I(\tau,\eta,\xi,\phi) = \frac{e^{\tau/\eta}}{\eta} \int_{\tau_0}^{\tau} B(\tau',\eta,\xi,\phi) e^{\tau'/\eta} \, d\tau' \text{ at } \eta < 0,$$

for the upward intensity, where we accounted for boundary conditions and $\eta = \cos\vartheta, \xi = \cos\vartheta_0$. Also, we define $\mu = |\eta|, \mu_0 = |\xi|$. It follows in terms of these angles: $\cos\theta = (-1)^l \mu\mu_0 + \sqrt{(1-\mu^2)(1-\mu_0^2)} \cos\phi$, where $l = 1$ for the reflected light and $l = 2$ for the transmitted light, respectively. If the function $B(\tau',\eta,\xi,\phi)$ is known, these two equations can be used to find the diffuse light field at any point inside the aerosol medium and also at its boundaries. The first boundary condition follows from the first equation assuming $\tau = 0$. The second boundary condition follows from the second equation assuming $\tau = \tau_0$. So both boundary conditions are satisfied.

Very often one needs to know the diffuse light field intensity escaping from the top $(I_\uparrow(0,\eta,\xi,\varphi))$ and the bottom $(I_\downarrow(\tau_0,\eta,\xi,\varphi))$ of a scattering layer. Their values can be obtained from two general equations given above. Then it follows that

$$I_\uparrow(0,\eta,\xi,\phi) = -\frac{1}{\eta} \int_0^{\tau_0} B(\tau',\eta,\xi,\phi) \exp\left(\frac{\tau'}{\eta}\right) d\tau', \, \eta < 0,$$

$$I_\downarrow(\tau_0, \eta, \xi, \phi) = \frac{\exp\left(-\frac{\tau_0}{\eta}\right)}{\eta} \int_0^{\tau_0} B(\tau', \eta, \xi, \phi) \exp\left(\frac{\tau'}{\eta}\right) d\tau', \quad \eta > 0,$$

where arrows identify the direction of propagation.

These equations allow for the following physical interpretation. Let us assume for the simplicity that $\eta = 1$. Then it follows:

$$I_\downarrow(\tau_0, \eta, \xi, \phi) = \int_0^{\tau_0} B(\tau', \eta, \xi, \phi) \exp(-(\tau_0 - \tau')) d\tau'.$$

Taking into account that the difference $\Delta \tau = \tau_0 - \tau'$ is equal to the optical thickness from the bottom of the layer to the level with the optical vertical coordinate τ', we conclude that the diffuse light intensity of the downward propagated light is determined by the summation of source functions $B(\tau', \eta, \xi, \phi)$ at levels $\tau = \tau'$ weighted by the exponential attenuation factors $\exp(-\Delta\tau)$ describing the attenuation of light from the level $\tau = \tau'$ to the lower boundary of an aerosol layer. In a similar way one can make an interpretation of the upward propagated intensity. Therefore, if the source function is known, the calculation of the diffuse intensity is straightforward. It is just reduced to the calculation of integrals from known functions, which can be done either analytically or numerically. This solves the mathematical problem at hand in a complete way. However, the trouble is that the derivation of the source function is by itself a complicated problem. For instance, one can obtain the following equation for the source function substituting I in the definition of B given above by just derived integrals. Then it follows that

$$B(\tau, \eta, \xi, \phi) = \frac{I_0 \omega_0}{4} p(\theta) e^{-\tau/\xi} + \frac{\omega_0}{4\pi} \int_0^{2\pi} d\phi' \left\{ \int_0^1 p(\theta') d\eta' \int_0^\tau B(\tau', \eta', \xi, \phi\prime) e^{(\tau'-\tau)/\eta'} \frac{d\tau'}{\eta'} - \right.$$

$$\left. - \int_{-1}^0 p(\theta') d\eta' \int_\tau^{\tau_0} B(\tau', \eta', \xi, \phi') e^{(\tau'-\tau)/\eta'} \frac{d\tau'}{\eta'} \right\}.$$

This equation cannot be solved analytically. Clearly, a numerical procedure is required. However, some simplifications arise in specific cases of large and small values of the aerosol optical thickness. For instance, it follows in the case of thin layers ($\tau_0 \to 0$) from the equation given above that

$$B(\tau, \eta, \xi, \phi) = \frac{I_0 \omega_0}{4} p(\theta) e^{-\tau/\xi}$$

and the analytical integration of corresponding integrals for the downward and upward intensities as presented above becomes possible. This is discussed in the next section.

3.3 Thin aerosol layers

For thin aerosol layers, all integrals in the integral equation for the source function can be neglected. The substitution of the remaining term $B = I_0 \omega_0 p(\theta) e^{-\tau/\xi}/4$ in the integrals for the diffuse light intensity gives for a homogeneous plane-parallel aerosol layer:

$$I_\uparrow(\tau, \vartheta, \vartheta_0, \phi) = \frac{\omega_0 p(\theta) I_0}{4(s-1)} [\exp(-x - (s-1)x_0) - \exp(-sx)],$$

$$I_\downarrow(\tau, \vartheta, \vartheta_0, \phi) = \frac{\omega_0 p(\theta) I_0}{4(s-1)} [\exp(-x) - \exp(-sx)],$$

where $\theta = \arccos(\cos\vartheta\cos\vartheta_0 + \sin\vartheta\sin\vartheta_0\cos\phi)$, ϕ is the relative azimuth of incident and diffused light beams, $s = \cos\vartheta/\cos\vartheta_0$, $x = \tau/\cos\vartheta$, $x_0 = \tau_0/\cos\vartheta$. These simple equations enable the calculation of the diffuse light intensity at any depth τ inside the scattering aerosol layer in any direction (ϑ, ϕ) for a given incidence angle ϑ_0 (see Fig. 3.1). As follows from Fig. 3.1, the value of ϑ varies from 0 to $\pi/2$ for transmitted light and from $\pi/2$ to π for reflected light. One derives for the light intensities at the boundaries of the aerosol layer:

$$I_\uparrow(0, \vartheta, \vartheta_0, \phi) = \frac{\omega_0 p(\theta) I_0}{4(s-1)} [\exp(-(s-1)x_0) - 1],$$

$$I_\downarrow(\tau_0, \vartheta, \vartheta_0, \phi) = \frac{\omega_0 p(\theta) I_0}{4(s-1)} [\exp(-x_0) - \exp(-sx_0)]$$

or

$$I_\uparrow(0, \mu, \mu_0, \phi) = \frac{\omega_0 p(\theta) \mu_0 I_0}{4(\mu_0 + \mu)} [1 - \exp(-\tau_0(\mu^{-1} + \mu_0^{-1}))],$$

$$I_\downarrow(\tau_0, \mu, \mu_0, \phi) = \frac{\omega_0 p(\theta) \mu_0 I_0}{4(\mu_0 - \mu)} [\exp(-\tau_0/\mu_0) - \exp(-\tau_0/\mu)],$$

where we introduced $\mu = |\eta|$, $\mu_0 = |\xi|$. It is of advantage to use characteristics normalized to the incident light flux in theoretical studies. The resulting equations represent the properties of the medium and they are invariant with respect to the incident flux. Corresponding characteristics and their integrals used in radiative transfer studies are given in Table 3.1. In particular, it follows for the reflection and transmission functions taking into account that the incident light flux E_0 is equal to πI_0:

$$R(\mu, \mu_0, \phi) = \frac{\omega_0 p(\theta)}{4(\mu_0 + \mu)} [1 - \exp(-\tau_0(\mu_0^{-1} + \mu^{-1}))],$$

$$T(\mu, \mu_0, \phi) = \frac{\omega_0 p(\theta)}{4(\mu_0 - \mu)} [\exp(-\tau_0/\mu_0) - \exp(-\tau_0/\mu)],$$

An important feature of these analytical equations is the fact that the reflection and transmission functions are symmetric with respect to the interchange of incidence and obser-

Table 3.1. Radiative transfer characteristics

Physical quantity	Symbol	Definition
Reflection function	$R(\mu_0,\mu,\phi)$	$\pi I_\uparrow(\mu_0,\mu,\phi)/\mu_0 E_0$
Plane albedo	$r_p(\mu_0)$	$\dfrac{1}{\pi}\displaystyle\int_0^{2\pi} d\phi \int_0^1 R(\mu_0,\mu,\phi)\mu\, d\mu$
Spherical albedo	r_s	$r_s = 2\displaystyle\int_0^1 r_p(\mu_0)\mu_0\, d\mu_0$
Transmission function	$T(\mu_0,\mu,\phi)$	$\pi I_\downarrow(\mu_0,\mu,\phi)/\mu_0 E_0$
Diffuse transmission	t_d	$\dfrac{1}{\pi}\displaystyle\int_0^{2\pi} d\phi \int_0^1 T(\mu_0,\mu,\phi)\mu\, d\mu$
Transmission coefficient	t	$t_s = 2\displaystyle\int_0^1 t_d(\mu_0)\mu_0\, d\mu_0$

vation directions. This is a manifestation of the general reciprocity principle (van de Hulst, 1980; Zege et al., 1991). It follows at $\mu = \mu_0$:

$$R = \frac{\omega_0}{8\mu_0}[1 - \exp(-2\tau_0)]p(\theta),$$

$$T = \frac{\omega_0 \tau_0}{4\mu\mu_0}\exp\left(-\frac{\tau_0}{\mu_0}\right)p(\theta).$$

One obtains as $\tau_0 \to 0$:

$$R = T = \frac{\omega_0 p(\theta)\tau_0}{4\mu_0^2}.$$

As $\tau_0 \to \infty$ it follows that $T = 0$ and

$$R = \frac{\omega_0 p(\theta)}{4(\mu + \xi)}.$$

This equation represents the contribution of the singly scattered light into the reflection function of a semi-infinite medium. One can see that this contribution is larger for weakly absorbing media ($\omega_0 \approx 1$) and large incidence and observation angles ($\mu \approx \xi \approx 0$ or $\vartheta \approx \vartheta_0 \approx \pi/2$). It depends on the phase function $p(\theta)$ of a scattering medium as well. As $\tau_0 \to 0$ one can obtain in the framework of the single scattering approximation that both R and T are determined by the same equation, e.g.,

$$R = \frac{\omega_0 \tau_0 p(\theta)}{4\mu_0\mu}.$$

One can see that functions

$$R^* = 4\mu_0\mu\, R, \quad T^* = 4\mu_0\mu T$$

do not depend on incidence and observation angles for isotropic scattering ($p \equiv 1$) as $\tau_0 \to 0$. They are just determined by the aerosol scattering depth $k_{sca}L \equiv \omega_0\tau_0$ in this case. Thus, the usage of the pair (R^*, T^*) has some advantages, as was pointed out by Chandrasekhar (1950). For anisotropic scattering, functions R^*, T^* (as $\tau_0 \to 0$) depend just on the scattering angle and not separately on the incidence, observation, and azimuth angles. In particular, it follows that

$$R^* = \omega_0\tau p(\theta)$$

or

$$R^* = \frac{4\pi}{k^2} N\bar{i}(\theta)L,$$

where we used relationships: $\omega_0 = k_{sca}/k_{ext}$, $\tau_0 = k_{ext}L$, $p(\theta) = 4\pi N\bar{i}(\theta)/k^2 k_{sca}$, $\bar{i}(\theta) = (\bar{i}_1(\theta) + \bar{i}_2(\theta))/2$. Here N is the number of particles in a unit volume and $\bar{i}_1(\theta)$, $\bar{i}_2(\theta)$ are the dimensionless Mie intensities averaged with respect to PSD (in the case of spherical polydispersions).

It follows for these functions that

$$R^* = \frac{4\pi\mu I_\uparrow^d}{E_0}, \quad T^* = \frac{4\pi\mu I_\downarrow^d}{E_0},$$

where E_0 is the incident light flux at the top of the layer (on the area perpendicular to the light beam).

The single scattering approximation is of importance for a number of reasons. First of all, the approximation can be used to calculate light intensity inside thin plane-parallel aerosol layers and also their reflective transmission characteristics using just Mie theory without the numerical solution of RTE. Secondly, this approximation is used as a building block in a number of exact techniques for the solution of RTE valid at any aerosol optical thickness. For instance, the adding–doubling method (van de Hulst, 1980) is based on the assumption that the light scattering by a very thin layer (e.g., $\tau_0 = 10^{-8}$) can be approximated by the equations given above. Then yet another layer is added at the top and the interaction between two layers is fully taken into account. This makes it possible to derive the result for the thicker layer. Clearly, this procedure can be applied many times to reach any required aerosol optical thickness. Yet another approximation is based on the account of the successive orders of scattering. The double scattering approximation is obtained by substituting the single scattering approximation for the source function in the integrand of the integral equation for the source function. This enables the derivation of the expression for the source function in the double scattering approximation. Subsequently, one can derive the diffuse intensity using the integral relationships between I and B presented above. These techniques can be used in addition to the currently most used approach based on the discretization of the integral term in the radiative transfer equation with subsequent solution of the system of differential equations (the method of discrete ordinates). A comparison of the main techniques for solving RTE and a comprehensive list of references is given by Lenoble et al. (1985). Currently, there are no difficulties related to the numerical solution of RTE and many codes are available online. They enable the determination of reflection and transmission functions and also the diffuse light field at any level inside the aerosol medium.

Clearly, the reflection function must increase with the thickness of the aerosol layer over a black surface (e.g., the ocean in IR). This feature is used for the determination of aerosol and cloud optical thickness from space.

There is an asymptotic value of the reflection function dependent on the incidence and observation angles, which is reached as $\tau_0 \rightarrow \infty$. This function is called the reflection function of a semi-infinite layer $R_\infty(\mu_0, \mu, \phi)$. Although aerosol layers are finite media, it is of importance to understand the properties of this function on general grounds. Also in some cases (e.g., fires, volcanic eruptions, and explosions) the optical thickness of an aerosol layer can indeed be very large and the reflection function is close to that of a semi-infinite layer then.

The calculations of $R_\infty(\mu_0, \mu, \phi)$ can be performed using, for example, the method of discrete ordinates and analyzing the results for the reflection function at a very large optical depth (e.g., 5000 at $\omega_0 = 1$ and much smaller values at small ω_0). However, the question arises: is it possible to formulate the radiative transfer equation for a semi-infinite layer in such a way that the radiative transfer equation does not contain the optical thickness at all. This will enable the most direct and accurate determination of $R_\infty(\mu_0, \mu, \phi)$ including all orders of scattering. The corresponding procedure was developed by Ambartsumian (1943). It is described in the next section.

3.4 Semi-infinite aerosol layers

The integral equation for the reflection function of a semi-infinite layer of a scattering medium can be derived using the principle of invariance developed by Ambartsumian (1943). The distinctive feature of this principle is the fact that it considers the properties of media as whole objects and does not use the consideration of energy balance for local processes, such as extinction, emission, and scattering similar to those described above, in the derivation of main equations.

Let us apply the principle of invariance to derive the integral equation for $R_\infty(\mu_0, \mu, \phi)$. The principle of invariance as applied to the problem at hand states that the reflection function of a semi-infinite layer does not change, if the layer with the optical thickness $\Delta\tau$ and the same values of ω_0, $p(\theta)$ as for a semi-infinite layer is added to the top of the semi-infinite scattering layer. This statement does not require any additional proof and in fact its representation in mathematical terms enables the derivation of the corresponding nonlinear integral equation for $R_\infty(\mu_0, \mu, \phi)$. The principle of invariance can be also applied to finite scattering layers. Then one must not only add a layer at the top of a medium but also subtract the same layer at the bottom. Clearly, the net effect of both operations must be equal to zero. This enables the derivation of important relationships for reflection and transmission functions of finite turbid layers (Ambartsumian, 1961). The general approach as introduced by Ambartsumian was later extended and applied in different branches of modern physics, astrophysics and mathematics (see, for example, Chandrasekhar, 1950; van de Hulst, 1980; Roth, 1986).

The corresponding integral equation derived using the invariance principles is written as (Ambartsumian, 1943):

$$R_\infty(\mu_0, \phi_0, \mu, \phi) =$$

$$\frac{\omega_0}{4(\mu + \mu_0)} p(\theta) + \frac{\mu_0 \omega_0}{4\pi(\mu_0 + \mu)} \int_0^1 \int_0^{2\pi} p(\mu, \phi, \mu', \phi') R_\infty(\mu', \phi', \mu_0, \phi_0) \, d\mu' \, d\phi'$$

$$+ \frac{\mu \omega_0}{4\pi(\mu_0 + \mu)} \int_0^1 \int_0^{2\pi} p(\mu_0, \phi_0, \mu', \phi') R_\infty(\mu', \phi', \mu, \phi) \, d\mu' \, d\phi'$$

$$+ \frac{\omega_0 \mu \mu_0}{4\pi^2(\mu_0 + \mu)} \int_0^{2\pi} d\phi' \int_0^1 R_\infty(\mu', \phi', \mu, \phi) \, d\mu' \int_0^{2\pi} d\phi'' \int_0^1 p(-\mu', \phi', \mu'', \phi'') R_\infty(\mu'', \phi'', \mu_0, \phi_0) \, d\mu'',$$

where θ is the scattering angle. This equation looks much more complicated as compared to standard RTE. However, in fact it can be easily solved numerically (see, for example, the code freely available at http://www.giss.nasa.gov/~crmim/ (Mishchenko et al., 1999)) and also leads to quick derivations of important analytical results. In particular, let us assume that the scattering is isotropic and, therefore, $p = 1$. Clearly, in this case the dependence on the azimuth disappears and it follows that

$$R_\infty(\mu, \mu_0) = \frac{\omega_0}{4(\mu + \mu_0)} [1 + \mu b(\mu) + \mu_0 b(\mu_0) + \mu \mu_0 b(\mu) b(\mu_0)],$$

where

$$b(\mu_0) = 2 \int_0^1 R_\infty(\mu_0, \mu) \, d\mu$$

is the plane albedo of a semi-infinite medium. The expression in brackets can be written as a product of functions $H(\mu)H(\mu_0)$, where $H(\mu) = 1 + \mu b(\mu)$. Therefore, one obtains:

$$R_\infty(\mu, \mu_0) = \frac{\omega_0 H(\mu) H(\mu_0)}{4(\mu + \mu_0)}.$$

This equation makes possible the reduction of the calculation of the reflection function of a semi-infinite layer with arbitrary absorption and $p(\theta) \equiv 1$ to the calculation of just one function of a single variable. This result is due to Ambartsumian (1961) and it demonstrates the power of the principle of invariance in the derivation of analytical solutions. This equation can be derived from the general RTE written at the beginning of this section as well (Chandrasekhar, 1950). However, the derivation presented here is simpler. Substituting the derived expression for $R_\infty(\mu, \mu_0)$ into the definition of $H(\mu)$, one derives the integral equation for the determination of the function $H(\mu)$:

$$H(\mu) = 1 + \frac{1}{2} \omega_0 \mu H(\mu) \int_0^1 \frac{H(\mu')}{\mu + \mu'} \, d\mu',$$

which can be solved analytically (Fock, 1944):

$$H(\mu) = \exp\left\{-\frac{\mu}{\pi}\int_0^\infty (1+\mu^2 x^2)^{-1}\ln\left(1 - \omega_0 \frac{\arctan x}{x}\right)dx\right\}.$$

It follows that $H(0) = 1$. Hapke (1993) proposed the following accurate approximation for this function:

$$H(\mu) = \frac{1 + 2\mu}{1 + 2\mu\sqrt{1 - \omega_0}},$$

which is valid with the error smaller than 4%. It follows from this approximation that

$$H(\mu) = 1 + 2\mu$$

at $\omega_0 = 1$. Therefore, one concludes that this function monotonically increases with μ from its value equal to one at $\mu = 0$ to $H = 3$ at $\mu = 1$ with $H(1/2) = 2$. The exact calculations give 1, 2.01, 2.91 at $\mu = 0$, 1/2, 1, respectively.

Light scattering by aerosols is not isotropic ($p \neq 1$). Therefore, the results presented above cannot be directly applied to aerosol optics problems. However, a number of approximations based on just the derived expression for the reflection function of a semi-infinite layer have been proposed (Hapke, 1993; Kokhanovsky, 2006). In particular, note that the reflection function of a semi-infinite layer having an arbitrary absorption can be represented as a series with respect to the number of scattering events. Clearly, the results for single scattering processes will be very different for isotropic and anisotropic scattering. However, for a large number of scatterings, the contribution of scattering events of a given order are less dependent on the phase function due to the property of multiple light scattering of washing out single scattering features. Therefore, a possible approximation for a semi-infinite layer of a scattering medium can be written as follows:

$$R_\infty(\mu,\mu_0,\phi) = \frac{\omega_0 H(\mu)H(\mu_0)}{4(\mu + \mu_0)} - \frac{\omega_0(1 - p(\theta))}{4(\mu + \mu_0)},$$

where we subtracted the contribution of the single isotropic light scattering and added the contribution of single anisotropic light scattering to the reflection function of a semi-infinite layer.

Therefore, it follows for the case of nonabsorbing media that

$$R_\infty^0(\mu,\mu_0,\phi) = \frac{p(\theta)}{4(\mu + \mu_0)} + \frac{H(\mu)H(\mu_0) - 1}{4(\mu + \mu_0)},$$

where $R_\infty^0(\mu,\mu_0,\phi)$ is the reflection function of a semi-infinite nonabsorbing medium with arbitrary phase function. The accuracy of this formula can be increased, if one does not use $H(\mu) \approx 1 + 2\mu$ valid for isotropic scattering only but rather finds the approximate fit for this function of a single argument from exact calculations of the difference

$$D(\mu,\mu_0) = R_\infty^0(\mu,\mu_0,\phi) - \frac{p(\theta)}{4(\mu + \mu_0)}.$$

In the case of a strongly absorbing semi-infinite aerosol layer, the quasi-single approximation holds. The corresponding analytical solution can be derived in the following way. We start from the general equation for the reflection function of a turbid plane-parallel homogeneous semi-infinite layer with an arbitrary single scattering albedo ω_0 and the phase function $p(\theta)$. This equation is given above and can be rewritten in the following form:

$$R_\infty(\mu_0,\mu,\phi,\phi_0) = \frac{\omega_0 p(\theta)}{4(\mu_0+\mu)} + \frac{\omega_0[\mu_0 V(\mu_0,\mu,\phi,\phi_0) + \mu V(\mu,\mu_0,\phi_0,\phi)]}{4\pi(\mu_0+\mu)} +$$
$$\frac{\omega_0\mu_0\mu W(\mu_0,\mu,\phi,\phi_0)}{4\pi^2(\mu_0+\mu)},$$

where

$$V(\mu_0,\mu,\phi,\phi_0) = \int\limits_0^{2\pi} d\phi' \int\limits_0^1 p(\mu,\phi,\eta',\phi')R(\mu_0,\eta',\phi',\phi_0)\,d\eta',$$

$$W(\mu_0,\mu,\phi,\phi_0) = \int\limits_0^{2\pi} d\phi' \int\limits_0^{2\pi} d\phi'' \int\limits_0^1 d\eta' \int\limits_0^1 R(\mu,\phi,\eta',\phi')p(-\eta',\phi',\eta'',\phi'')R(\mu_0,\eta'',\phi'',\phi_0)\,d\eta''.$$

The numerical solution of this equation can be used to check the accuracy of various approximations.

It follows that the reflection function can be presented as

$$R_\infty(\mu_0,\mu,\phi,\phi_0) = \frac{\omega_0 p(\theta)}{4(\mu_0+\mu)} + \Pi,$$

where the first term accounts for single scattering and the second gives the contribution of multiple light scattering:

$$\Pi = \frac{\omega_0[\mu_0 V(\mu_0,\mu,\phi,\phi_0) + \eta V(\mu,\mu_0,\phi_0,\phi)]}{4\pi(\mu_0+\mu)} + \frac{\omega_0\mu_0\mu W(\mu_0,\mu,\phi,\phi_0)}{4\pi^2(\mu_0+\mu)}.$$

The integral V can be simplified for media with highly anisotropic scattering. Then phase functions are highly extended in the forward direction. This means that we assume that $p(\mu,\phi,\eta',\varphi')$ is close to the delta function. Then it follows, using the definition of delta function, that

$$V(\mu_0,\mu,\phi,\phi_0) = \int\limits_0^{2\pi} d\phi' \int\limits_0^1 p(\mu,\phi,\eta',\varphi')R(\mu_0,\eta',\varphi',\phi_0)\,d\eta'$$

$$\approx R(\mu_0,\mu,\phi,\phi_0)\int\limits_0^{2\pi} d\varphi' \int\limits_0^1 p(\mu,\phi,\eta',\varphi')\,d\eta' = 4\pi R(\mu_0,\mu,\phi,\phi_0)c(\mu,\phi_0),$$

where

$$c(\mu,\phi) = \frac{1}{4\pi} \int\limits_0^{2\pi} d\phi' \int\limits_0^1 p(\mu,\phi,\eta',\phi')\,d\eta'.$$

The phase function depends only on the difference $\varphi = \phi - \phi_0$. This means that

$$c(\mu) = \frac{1}{4\pi} \int_0^{2\pi} d\phi \int_0^1 p(\mu, \eta', \phi)\, d\eta'.$$

So we have:

$$R_\infty(\mu_0, \mu, \varphi) = \frac{\omega_0 p(\theta)}{4(\mu_0 + \mu)(1 - \omega_0 F^*)},$$

where

$$F^* = \frac{\mu_0 c(\mu) + \mu c(\mu_0)}{\mu_0 + \mu}.$$

This formula was derived by Anikonov and Ermolaev (1975). They also have shown that $c(\mu)$ can be reduced to the following simpler form:

$$c(\mu) = F - \Delta(\mu),$$

where

$$F = \frac{1}{2} \int_0^1 p(\eta)\, d\eta$$

$$\Delta(\mu) = \frac{1}{2\pi} \int_0^{\sqrt{1-\mu^2}} d\eta \{p(\eta) - p(-\eta)\} \arccos\left\{ \frac{\mu\eta}{\sqrt{(1-\mu^2)(1-\eta^2)}} \right\}.$$

Our numerical calculations show that $\Delta(\mu) \ll F$. Therefore, we can neglect the contribution of Δ. This reduces the expression for the reflection function of a semi-infinite strongly absorbing aerosol layer to the simpler formula known as the Gordon approximation (Gordon, 1973):

$$R_\infty(\mu_0, \mu, \phi) = \frac{\omega_0 p(\theta)}{4(\mu_0 + \mu)(1 - \omega_0 F)}.$$

One concludes that the reflection function in the framework of the considered approximation is just equal to the product of the contribution of the single light scattering

$$R_{ss\infty} = \frac{\omega_0 p(\theta)}{4(\mu_0 + \mu)}$$

and the factor $\Im = (1 - \omega_0 F)^{-1}$, which accounts for the effects of multiple light scattering.

3.5 Thick aerosol layers

One might expect that the results just derived for a semi-infinite light scattering medium can also be used with a slight modification in the case of large but not infinite values of τ_0 (i.e., in the so-called asymptotic regime, where $\tau_0 \gg 1$). This is indeed the case, as demonstrated by Germogenova (1961) and van de Hulst (1980). The final expressions for the reflection and transmission functions take the following forms valid at any ω_0 and $p(\theta)$ assuming that the aerosol optical thickness is larger than at least 5:

$$R(\mu_0, \mu, \phi) = R_\infty(\mu_0, \mu, \phi) - T(\mu_0, \mu) l\, e^{-k\tau},$$

$$T(\mu_0, \mu) = \frac{m\, e^{-k\tau}}{1 - l^2\, e^{-2k\tau}} K(\mu_0) K(\mu).$$

Here the pair (μ_0, μ) gives the cosines of the incidence and observation angles, ϕ is the relative azimuth. $R_\infty(\mu_0, \mu, \phi)$ is the reflection function of a semi-infinite scattering layer having the same local optical characteristics (e.g., the same single scattering albedo ω_0 and the same phase function $p(\theta)$ with scattering angle θ) as the finite layer currently under study. The constants (k, l, m) and the escape function $K(\mu)$ do not depend on τ and can be obtained from the solution of integral equations as described by van de Hulst (1980), Wauben (1992), and Kokhanovsky (2006).

For the use of the analytical equations given above at arbitrary ω_0 and $p(\theta)$, several parameters $(k, l, m, n, r_{s\infty})$ and also functions $K(\mu)$, $R_\infty(\mu_0, \mu, \phi)$ and $r_{p\infty}(\mu)$ have to be derived (Kokhanovsky, 2004b, 2006). The problem is simplified at $\omega_0 = 1$. Then it follows: $k = m = 0$, $l = 1$. Parameters k, l, m can be parameterized as follows at arbitrary values of single light scattering albedo (King and Harshvardan, 1986):

$$k = \left(\sqrt{3}s - \frac{(0.985 - 0.253s)s^2}{6.464 - 5.464s} \right)(1 - \omega_0 g),$$

$$l = \frac{(1-s)(1 - 0.681s)}{1 + 0.729s},$$

$$m = (1 + 1.537s) \ln\left(\frac{1 + 1.8s - 7.087s^2 + 4.74s^3}{(1 - 0.819s)(1 - s)^2} \right),$$

where

$$s = \sqrt{\frac{1 - \omega_0}{1 - \omega_0 g}}$$

is the similarity parameter and

$$g = \frac{1}{4} \int_0^\pi p(\theta) \sin(2\theta)\, d\theta$$

is the asymmetry parameter.

Functions $K(\mu)$ and $R_\infty(\mu_0, \mu, \phi)$ cannot be parameterized in terms of the similarity parameter alone. Therefore look-up-tables (LUTs) of these functions can be used (Kokhanovsky and Nauss, 2006). For nonabsorbing media, the following approximation holds at $\mu \geq 0.2$:

$$K(\mu) = \frac{3}{7}(1 + 2\mu).$$

Approximate equations for the function $R_\infty(\mu_0, \mu, \phi)$ are presented in the previous section. Kokhanovsky and Nauss (2006) developed a numerical code based on equations specified above and LUTs for $K(\mu)$ and $R_\infty(\mu_0, \mu, \phi)$. The code is freely available at the website www.iup.physik.uni-bremen.de/~alexk.

3.6 Aerosols over reflective surfaces

The results given above can be easily generalized to account for the light reflection from the aerosol layer with an underlying Lambertian surface. Let us derive corresponding equations. Light intensity observed in the direction specified by the pair (ϑ, φ) can be considered as composed of two parts: that due to an aerosol layer itself (I_1) and that due to the surface contribution (I_2). The contribution I_2 can be also separated into two terms (I_{21}, I_{22}), namely

$$I_{21} = I_s t(\mu)$$

for the contributions of the surface in the diffused light and

$$I_{22} = I_s \exp(-\tau/\mu)$$

for the contribution of the surface in the direct light. Here $t(\mu)$ is the diffuse transmittance defined as

$$t(\mu) = 2 \int_0^1 \overline{T}(\mu_0, \mu)\mu_0 \, d\mu_0,$$

where

$$\overline{T}(\mu_0, \mu) = \frac{1}{2\pi} \int_0^{2\pi} T(\mu_0, \mu, \phi) \, d\phi.$$

Summing up, we have:

$$I(\mu, \varphi) = I_1(\mu, \varphi) + I_s t(\mu) + I_s \, e^{-\tau/\mu},$$

where we assumed that the surface is a Lambertian reflector. This means that the upward intensity I_s for the light emerging from the ground surface does not depend on the angle. Let us relate I_s to the albedo A of underlying Lambertian surface. For this we note that the upward flux density is

$$F_u = \int_{2\pi}^{I_s} \cos \vartheta \, d\Omega = \int_0^{2\pi} d\varphi \int_0^{\pi/2} d\vartheta I_s \cos \vartheta \sin \vartheta = \pi I_s.$$

We have for the ideally reflecting Lambertian surface ($A = 1$): $F_u = F_d$, where F_d is the downward flux density. F_d is composed of three components: direct transmission component $F_{dir} = \mu_0 F_0 e^{-\tau/\mu_0}$, the diffused transmission component $F_{dif} = \mu_0 F_0 t(\mu_0)$ and the component coming from the surface but reflected by a scattering layer back to the underlying surface: $F_{ref} = r F_u$, where r is the spherical albedo of a scattering aerosol layer under illumination from below defined as

$$r = 2 \int_0^1 r_p(\mu) \mu \, d\mu,$$

where

$$r_p(\mu) = 2 \int_0^1 \overline{R}(\mu_0, \mu) \mu_0 \, d\mu_0,$$

$$\overline{R}(\mu_0, \mu) = \frac{1}{2\pi} \int_0^{2\pi} R(\mu_0, \mu, \phi) \, d\phi.$$

Here $r_p(\mu)$ is the plane albedo. Obviously, for the underlying surface with an arbitrary ground albedo A, we have:

$$F_u = A F_d$$

and, therefore,

$$\pi I_s = A \left[\mu_0 F_0 \left(t(\mu_0) + e^{-\tau/\mu_0} \right) + \pi r I_s \right]$$

The intensity I_s can be easily found from this equation. It follows that

$$I_s = \frac{A t^*(\mu_0) \mu_0 F_0}{\pi (1 - Ar)}$$

where

$$t^*(\mu_0) = t(\mu_0) + e^{-\tau/\mu_0}$$

is the total transmittance. Therefore, we have (Liou, 2002):

$$I(\mu, \varphi) = I_1(\mu, \varphi) + \frac{A t^*(\mu) t^*(\mu_0) \mu_0 F_0}{\pi (1 - Ar)}$$

or

$$R(\mu_0, \mu, \varphi) = R_b(\mu_0, \mu, \varphi) + \frac{A t^*(\mu_0) t^*(\mu)}{1 - Ar},$$

where $R_b(\mu_0, \mu, \varphi) \equiv R(\mu_0, \mu, \varphi)$ at $A = 0$. All functions presented in this equation have been studied in the previous section.

A similar simple account for the Lambertian underlying surface can also be performed for the transmitted component. Namely, we have then:

$$I_{tr}(\mu,\varphi) = I_{1tr}(\mu,\varphi) + I_s r_p(\mu),$$

where the first component is due to light transmission by the aerosol layer itself and the second component accounts for the reflection of the diffused light (I_s) coming from the surface, $r_p(\mu)$ is the plane albedo illumination from below. Finally, one derives:

$$I_{tr}(\mu,\varphi) = I_{1tr}(\mu,\varphi) + \frac{At^*(\mu_0)r_p(\mu)\mu_0 F_0}{\pi(1 - Ar)}$$

or, for the transmission function,

$$T(\mu_0,\mu,\varphi) = T_b(\mu_0,\mu,\varphi) + \frac{Ar_p(\mu_0)t^*(\eta)}{1 - Ar},$$

where $T_b(\mu_0,\mu,\varphi) \equiv T(\mu_0,\mu,\varphi)$ at $A = 0$.

The parameterizations of the function $t(\eta)$ and also r in terms of the aerosol optical thickness and the asymmetry parameter g have been proposed by several authors (see, for example, Kokhanovsky et al., 2005). Such parameterizations are useful in satellite aerosol retrieval algorithms.

3.7 Multiple scattering of polarized light in aerosol media

3.7.1 The vector radiative transfer equation and its numerical solution

Light coming to the Earth from the Sun is unpolarized. However, it becomes polarized due to interaction with molecules and particles present in the terrestrial atmosphere. In particular, the theory of molecular scattering (Rayleigh, 1871) states that the polarization of initially unpolarized light after single scattering by a unit volume of air is almost 100 % at right angles to the direction of incidence. Macroscopic particles such as dust grains and ice crystals have smaller polarization ability at a scattering angle of 90 degrees (see, e.g., Fig. 2.14(a)). However, they also polarize light and can produce quite large values of the degree of polarization in some selected directions.

It is a well known fact that calculations of scattered light intensity I (both in single and multiple light scattering regimes) cannot be done accurately without accounting for the polarization characteristics of a light beam (Rozenberg, 1955; Hovenier, 1971; Mishchenko and Travis, 1997; Lacis et al., 1998; Mishchenko et al., 2002; Min and Duan, 2004). It means that the vector radiative transfer equation (VRTE) should be used whenever it is possible for studies of light transport in the atmosphere and other turbid media. The use of the scalar radiative transfer equation (SRTE) studied above could lead to errors in many cases. Also the solution of the VRTE is of importance for remote sensing techniques (Hansen and Hovenier, 1974; Hansen and Travis, 1974; de Haan, 1987; Goloub et al., 2000; Mishchenko and Travis, 1997; Mishchenko et al., 2002; Kokhanovsky, 2003, 2004a). In particular, optical instruments are capable of measuring not only I but also other components of the Stokes vector $\vec{S}(I,Q,U,V)$ (Deschamps et al., 1994). Spectral and angular measurements of $\vec{S}$ bring us much more information on the medium under study as compared to just light intensity I (Kokhanovsky, 2003). However, the usage of polarized light in astronomical applications, remote sensing and optical diagnostics of various turbid media is not widespread so far, owing to the complexity of the VRTE for the Stokes vector $\vec{S}$. Also the corresponding optical instruments are more complex because they measure simultaneously not only the intensity but also three additional parameters, which characterize the polarization properties of the light beam. Finding the numerical solution of this integro-differential vector radiative transfer equation is quite a complex mathematical procedure (Hovenier, 1971; Siewert, 2000). All the approximate and numerical techniques described in the previous section designed at first for the solution of SRTE have been generalized and used to solve VRTE. The main results in this direction have been reviewed by Kokhanovsky (2003, 2006), Hovenier et al. (2004) and Mishchenko et al. (2006).

To avoid the repetition of theories described above for the case of polarized light, which in many respects is very similar to the scalar case (see, for example, Kokhanovsky, 2003) the discrete ordinate technique (DOT) is considered in detail here. DOT was mentioned above but has not been described in detail so far. The main results with respect to DOT were obtained by Chandrasekhar (1950). The numerical codes and details of the method are given by Thomas and Stamnes (1999) and Siewert (2000) among others. The superiority of DOT over the doubling–adding (de Haan, 1987) and Monte Carlo (Tynes et al., 2001, Ishimoto and Masuda, 2002) methods is due to the weak dependence of the speed of computer simulations on the optical thickness. Clearly, the results described below can be used for the solution of the SRTE as well. Then one must use just the first equation for the diffuse light intensity and, correspondingly, just the first element of the phase matrix, which in fact coincides with the phase function.

The vector radiative transfer equation for the total Stokes vector of light beam propagating in a homogeneous isotropic symmetric plane-parallel light scattering aerosol medium is usually written as (Siewert, 2000):

$$\mu \frac{d\vec{S}(\tau,\mu,\phi)}{d\tau} = -\vec{S}(\tau,\mu,\phi) + \frac{\omega_0}{4\pi} \int\limits_{-1}^{1} d\mu' \int\limits_{0}^{2\pi} d\phi' \hat{L}(\alpha)\hat{P}(\mu,\mu',\phi-\phi')\hat{L}(\beta)\vec{S}(\tau,\mu',\phi')$$

for $\tau \in [0,\tau_0], \mu \in [-1,1]$ and $\phi \in [0,2\pi]$. Here τ_0 is the optical thickness, ω_0 is the single scattering albedo, $\hat{P}(\mu,\mu',\phi-\phi')$ is the phase matrix in the coordinate system attached to the scattering plane, μ is the cosine of the polar angle ϑ as measured from the positive τ-axis and ϕ is the azimuthal angle. The value of τ is the optical depth changing from 0 at the top of the turbid plane-parallel layer to the value $\tau = \tau_0$ at the bottom.

The components of the Stokes vector $\vec{S}$ are defined as follows (van de Hulst, 1980): $I = I_1 + I_r, Q = I_1 - I_r, U = E_1 E_r^* + E_r E_1^*, V = i(E_1 E_r^* - E_r E_1^*)$, where we neglect a common multiplier and $I_1 = E_1 E_1^*$ is the scattered light intensity in the meridional plane. This plane contains the normal to a light scattering slab and the direction of observation. The value of $I_r = E_r E_r^*$ gives the scattered light intensity in the plane perpendicular to the meridional plane. E_1 and E_r are components of the electric vector of the scattered wave defined relatively to the meridional plane in the same way as I_1, I_r (van de Hulst, 1980).

The phase matrix $\hat{P}(\mu,\mu',\phi-\phi')$ is defined with respect to the scattering plane containing incident and scattered light beams. Therefore, the Stokes vector $\vec{S}$ defined with respect to the meridional plane must be rotated using the rotation matrix $\hat{L}(\beta)$ (see the VRTE given above). This makes it possible to apply the phase matrix to the rotated Stokes vector

$\vec{S}' = \hat{L}(\beta)\vec{S}$ defined in the scattering plane. The second rotation is needed to bring the scattered Stokes vector $\vec{S}_{\text{sca}} = \hat{P}\hat{L}(\beta)\vec{S}$ back to the meridional plane. Hence, the scattered Stokes vector in the meridional plane is given by the product $\hat{L}(a)\hat{P}\hat{L}(\beta)\vec{S}$. This explains the rationale behind the appearance of rotation matrices in the VRTE. The rotation matrix for the Stokes vector has the following standard form (Mishchenko et al., 2002):

$$\hat{L}(\varphi) = \begin{pmatrix} 1 & 0 & 0 & 0 \\ 0 & \cos 2\varphi & -\sin 2\varphi & 0 \\ 0 & \sin 2\varphi & \cos 2\varphi & 0 \\ 0 & 0 & 0 & 1 \end{pmatrix},$$

if the rotation through the angle φ in the clockwise direction when looking in the direction of propagation is performed. The spherical trigonometry gives the following relationships between the pairs $(\cos a', \cos \beta')$ and (μ, μ'), where we introduced the angles $a' = -a$ and $\beta' = \pi - \beta$ to have correspondent equations in the symmetric form:

$$\cos a' = \frac{\mu' - \mu \cos \theta}{\sqrt{1 - \mu^2} \sin \theta}, \quad \cos \beta' = \frac{\mu - \mu' \cos \theta}{\sqrt{1 - \mu'^2} \sin \theta}.$$

The scattering angle is defined as introduced in the previous chapter:

$$\cos \theta = \mu\mu' + \sqrt{(1 - \mu^2)(1 - \mu'^2)} \cos \phi.$$

The necessity to perform rotations complicates the corresponding theory, especially in the case if anisotropic aerosol media are under consideration. This can be avoided if one uses the formulation of the RTE (Kokhanovsky, 2003) in the tensor form invariant with respect to the choice of the coordinate system.

We will assume that there is a Lambertian surface with the spherical albedo A underlying a plane-parallel aerosol layer. It is also assumed that the optical properties of the medium are the same at any point $M(\vec{r})$ inside a slab. The slab is illuminated by a wide unidirectional light beam at the top ($\tau = 0$). Both medium and light source are assumed to be time-independent and possible nonlinear and close-packed effects are neglected. The task is to find the vector $\vec{S}$ at any point M with the radius-vector $\vec{r}$ inside and outside of the scattering medium for arbitrary values of ω_0, τ_0 and phase matrices $\hat{P}(\mu, \mu', \phi - \phi')$. The main steps of the discrete ordinate technique to solve the VRTE are outlined below.

Step 1. The rotated phase matrix $\hat{P}^* = \hat{L}(a)\hat{P}\hat{L}(\beta)$ is presented in the form:

$$\hat{P}^*(\mu, \mu', \phi - \phi') = \frac{1}{2} \sum_{m=0}^{N} (2 - \delta_{0m}) \left\{ \begin{array}{l} \langle \hat{A}^m(\mu, \mu') + \hat{D}\hat{A}^m(\mu, \mu')\hat{D} \rangle \cos(m(\phi - \phi')) + \\ \langle \hat{A}^m(\mu, \mu')\hat{D} - \hat{D}\hat{A}^m(\mu, \mu') \rangle \sin(m(\phi - \phi')) \end{array} \right\}.$$

where $\hat{A}^m = \sum_{l=m}^{N} \hat{P}_l^m(\mu)\hat{\Xi}_l\hat{P}_l^m(\mu')$, $\hat{D} = \text{diag}\{1, 1, -1, -1\}$, δ is the Kronecker symbol, $\mathbb{N}$ is the maximal order of Legendre polynomials used, and

$$\hat{\Xi}_l = \begin{pmatrix} a_{11} & \beta_{11} & 0 & 0 \\ \beta_{11} & a_{21} & 0 & 0 \\ 0 & 0 & a_{31} & \beta_{21} \\ 0 & 0 & -\beta_{21} & a_{41} \end{pmatrix},$$

$$\hat{P}_1^m(\mu) = \begin{pmatrix} P_1^m(\mu) & 0 & 0 & 0 \\ 0 & -\frac{1}{2}i^m\left\langle P_{m,2}^l(\mu)+P_{m,-2}^l(\mu)\right\rangle & \frac{1}{2}i^m\left\langle P_{m,2}^l(\mu)-P_{m,-2}^l(\mu)\right\rangle & 0 \\ 0 & \frac{1}{2}i^m\left\langle P_{m,2}^l(\mu)-P_{m,-2}^l(\mu)\right\rangle & -\frac{1}{2}i^m\left\langle P_{m,2}^l(\mu)+P_{m,-2}^l(\mu)\right\rangle & 0 \\ 0 & 0 & 0 & P_1^m(\mu) \end{pmatrix}$$

with

$$P_1^m(\mu) = \sqrt{\frac{(l-m)!}{(l+m)!}}\{1-\mu^2\}^{m/2}\frac{d^m}{d\mu^m}P_1(\mu),$$

$$P_{m,n}^l(\mu) = \frac{(-1)^{l-m}i^{n-m}}{2^l(l-m)!}\sqrt{\frac{(l-m)!(l+n)!}{(l+m)!(l-n)!}}\{1-\mu\}^{(m-n)/2}\{1+\mu\}^{(m+n)/2}\frac{d^{l-n}}{d\mu^{l-n}}\left[\{1-\mu\}^{l-m}\{1+\mu\}^{l+m}\right].$$

Here

$$P_1(\mu) = \frac{1}{2^l l!}\frac{d^l}{d\mu^l}(\mu^2-1)^l$$

are Legendre polynomials, $P_1^m(\mu)$ are associated Legendre functions, $P_{m,n}^l(\mu)$ are generalized spherical functions. The Greek constants $\{a_{11}, a_{21}, a_{31}, a_{41}, \beta_{11}, \beta_{21}\}$ are determined by a local scattering law. For instance, it follows for the dipole scattering (de Rooij, 1985) that

$$a_{10} = 1, a_{12} = \frac{1}{2}, a_{22} = 3, a_{41} = \frac{3}{2}, \beta_{12} = \sqrt{\frac{3}{2}}$$

with all other constants being equal to zero. The table of Greek constants for media composed of identical randomly oriented oblate spheroids with the aspect ratio 2, the size parameter 3 and the refractive index 1.53–0.006i is given by Kuik et al. (1992). Corresponding constants can be easily obtained for monodispersed and polydispersed spherical particles as well. Then the Mie theory (van de Hulst, 1957) can be used (see, for example, the FORTRAN code *spher.f* located at http://www.giss.nasa.gov/~crmim/brf).

Although the formulation presented above looks quite cumbersome, it enables the substitution of the rotated phase matrix $\hat{P}$ by the discrete Greek symbols. These symbols can be used to find $\hat{P}^*$ at any combination of μ, μ', and $\phi = \varphi - \varphi'$. The algorithms of finding matrices $\hat{P}_1^m(\mu)$ and $\hat{\Xi}_1$ are well known and straightforward (de Rooij, 1985; Siewert, 1997; Mishchenko et al., 2002). In particular, for calculations of generalized spherical functions one can use Wigner functions $d_{m,\pm2}^l(\vartheta)$ (Mishchenko et al., 2002):

$$-\frac{1}{2}i^m\left\langle P_{m,2}^l(\mu)+P_{m,-2}^l(\mu)\right\rangle = \frac{1}{2}(-1)^m\left\langle d_{m,2}^l(\vartheta)+d_{m,-2}^l(\vartheta)\right\rangle.$$

This makes it possible to avoid calculations involving complex functions and use stable and accurate algorithms to calculate Wigner functions as described by Mishchenko et al. (2002). One can use properties:

$$P_1^m(\mu) = \sqrt{\frac{(l+m)!}{(l-m)!}}d_{m,0}^l(\vartheta)$$

to calculate the associated Legendre function.

The microstructure of the aerosol medium (e.g., the size and shape distributions, refractive indices of particles) enters the theory only via Greek symbols specified above. Greek symbols are determined by the phase matrix $\hat{P}$ of a single scattering law. The $4*4$ matrix $\hat{P}$ is defined with respect to the scattering plane holding directions of incident and scattered beams. Due to the symmetry of the media under consideration, elements of this matrix correspondent to upper-right and down-left $2*2$ sub-matrices vanish. Greek symbols can be calculated from the elements of the matrix $\hat{P}$ using following equations (de Rooij, 1985):

$$a_{11} = \frac{2l+1}{2} \int_{-1}^{1} dx P_l^{00}(x) P_{11}(x), \quad a_{41} = \frac{2l+1}{2} \int_{-1}^{1} dx P_l^{00}(x) P_{44}(x) \quad ,$$

$$\beta_{11} = \frac{2l+1}{2} \int_{-1}^{1} dx P_l^{02}(x) P_{12}(x), \quad \beta_{21} = \frac{2l+1}{2} \int_{-1}^{1} dx P_l^{02}(x) P_{34}(x),$$

and $a_{21} = \frac{1}{2}\{v_l + \varsigma_l\}$, $a_{31} = \frac{1}{2}\{v_l - \varsigma_l\}$, where $P_l^{s,m}(x) = i^s \sqrt{(1+|s|!)/(1-|s|!)} P_{s,m}^l(x)$,

$$v_l = -\frac{2l+1}{2} \sqrt{\frac{(l-2)!}{(l+2)!}} \int_{-1}^{1} dx P_l^{2,-2}(x)\{P_{22}(x) - P_{33}(x)\},$$

$$\varsigma_l = -\frac{2l+1}{2} \sqrt{\frac{(l-2)!}{(l+2)!}} \int_{-1}^{1} dx P_l^{2,-2}(x)\{P_{22}(x) + P_{33}(x)\}.$$

Step 2. The following Fourier expansion of the Stokes vector is used:

$$\vec{S} = \frac{1}{2} \sum_{m=0}^{L} \sum_{k=1}^{2} \hat{\Phi}_k^m(\phi - \phi_0) \vec{S}_k^m(\tau, \mu) + \pi \delta(\mu - \mu_0) \delta(\phi - \phi_0) \vec{I}_0 \exp(-\tau/\mu),$$

where $\mu_0 = \cos \vartheta_0$, ϑ_0 is the solar zenith angle, ϕ_0 is the solar azimuth, and $\delta(\varsigma - \varsigma_0)$ is the delta function. The second term explicitly accounts for the attenuated direct light ($\vec{I}_0$ is the Stokes vector of the incident light flux) and (Siewert, 2000)

$$\hat{\Phi}_1^m(\phi) = (2 - \delta_{0m}) \, \text{diag}\{\cos m\phi, \cos m\phi, \sin m\phi, \sin m\phi\},$$

$$\hat{\Phi}_2^m(\phi) = (2 - \delta_{0m}) \, \text{diag}\{-\sin m\phi, -\sin m\phi, \cos m\phi, \cos m\phi\}.$$

Therefore, the components $\vec{S}_k^m(\tau, \mu)$ describe not the total but just the diffuse light field. A similar separation of the attenuated direct and diffuse light is used in the scalar radiative transfer theory. Actually, all results of the scalar radiative transfer theory are obtained from corresponding results of the vector theory, if the Stokes vector is reduced to just its first component and the phase matrix is reduced to the phase function.

Step 3. The substitution of expressions for the Stokes vector and also for the rotated phase matrix in terms of series in the VRTE gives:

$$\mu \frac{d\vec{S}_k^m(\tau,\mu)}{d\tau} = -\vec{S}_k^m(\tau,\mu) + \frac{\omega_0}{2} \sum_{l=m}^{L} \hat{P}_l^m(\mu)\hat{\Xi}_l \int_{-1}^{1} d\mu' \hat{P}_l^m(\mu')\vec{S}_k^m(\tau,\mu') + \vec{Q}_k^m(\tau,\mu),$$

where

$$\vec{Q}_k^m(\tau,\mu) = \frac{\omega_0}{2} \sum_{l=m}^{L} \hat{P}_l^m(\mu)\hat{\Xi}_l \hat{P}_l^m(\mu_0)\hat{D}_k \vec{I}_0 \exp(-\tau/\mu_0),$$

and $\hat{D}_1 = \mathrm{diag}\{1,1,0,0\}$, $\hat{D}_2 = \mathrm{diag}\{0,0,1,1\}$. For the simplicity, we will consider only the case of unpolarized incident light. Then it follows that $\vec{I} = I_0(1,0,0,0)^\mathrm{T}$, where $\pi\mu_0 I_0$ is equal to the incident light irradiance at the top of the layer. This equation is much simpler to handle than the general VRTE given in the beginning of this chapter because the integration with respect to the azimuthal angle was performed analytically. The single integral can be substituted by series leading to the system of differential equations:

$$\pm\mu_\sigma \frac{d\vec{S}(\tau,\pm\mu_\sigma)}{d\tau} = -\vec{S}(\tau,\pm\mu_\sigma) + \frac{\omega_0}{2} \sum_{l=m}^{L} \hat{P}_l^m(\pm\mu_\sigma)\hat{\Xi}_l \sum_{\rho=1}^{N} w_\rho \vec{S}_{1,\rho}(\tau) + \vec{Q}(\tau,\pm\mu_\sigma),$$

where $\sigma = 1, 2, \ldots N$, N is the number of Gauss quadrature points, $\vec{S}_{1,\rho}(\tau) = \hat{P}_l^m(\mu_\rho)\vec{S}(\tau,\mu_\rho) + \hat{P}_l^m(-\mu_\rho)\vec{S}(\tau,-\mu_\rho)$ and the Fourier indices are suppressed for the sake of simplicity. Gauss quadrature points $\{\mu_\sigma\}$ and weights $\{w_\rho\}$ in this equation are defined for the use on the integration interval $[0,1]$. For unpolarized incident light illumination conditions we have : $k = 1$. So the corresponding index is omitted. Therefore, we conclude that the system of four integro-differential equations is substituted by the system of differential equations (SDE), which is simple to solve, using, for example, standard routines.

Step 4. The system of differential equations can be solved using the DOT as described by Siewert (2000). In particular, the solution of the homogeneous equation (with $\vec{Q} = 0$ in the SDE, see above) is found as follows:

$$\vec{S}_\pm^h(\tau) = \vec{R}_\pm(\tau) + \vec{C}_\pm(\tau),$$

where

$$\vec{R}_\pm(\tau) = \Delta_\pm \sum_{j=1}^{J_R} \left[A_j \vec{\Phi}_\pm(v_j) \exp\left\{-\frac{\tau}{v_j}\right\}^+ B_j \vec{\Phi}_\mp(v_j) \exp\left\{-\frac{\tau_0 - \tau}{v_j}\right\}\right],$$

$$\vec{C}_\pm(\tau) = \Delta_\pm \sum_{\upsilon=1}^{2} \sum_{j=1}^{J_C} \left[A_j^{(\upsilon)} \vec{F}_\pm^{(\upsilon)}(\tau, v_j) + B_j^{(\upsilon)} \vec{F}_\mp^{(\upsilon)}(\tau_0 - \tau, v_j)\right],$$

$$\vec{F}_\pm^{(1)}(v_j) = \mathrm{Re}\left\{\vec{\Phi}_\pm(v_j) \exp\left\{-\frac{\tau}{v_j}\right\}\right\}, \quad \vec{F}_\pm^{(2)}(v_j) = \mathrm{Im}\left\{\vec{\Phi}_\pm(v_j) \exp\left\{-\frac{\tau}{v_j}\right\}\right\}.$$

Here $\Delta_+ = \mathrm{diag}\{1, 1, \ldots, 1\}, \Delta_- = \mathrm{diag}\{\hat{D}, \hat{D}, \ldots, \hat{D}\}$ are $4N * 4N$ diagonal matrices, v_j is the collection of separation constants, J_R is the number of real eigenvalues, J_C is the number of complex eigenvalues. Vectors $\vec{\Phi}_{\pm}(v_j)$ with appropriate exponential multipliers are elementary solutions of a homogeneous equation, which can be found after solution of the corresponding eigenvalue problem (Rozanov and Kokhanovsky, 2006).

The particular solution of the inhomogeneous SDE is found using the infinite-medium Green function approach. Correspondent derivations are given by Siewert (2000) and the result is:

$$\vec{S}_{\pm}^p(\tau) = \Delta_\pm \sum_{j=1}^{J_R} \left[A_{(\tau)} \vec{\Phi}_{\pm}(v_j)^+ B_j(\tau) \vec{\Phi}_{\mp}(v_j) \right]$$

$$+ 2\Delta_\pm \, \mathrm{Re} \sum_{j=1}^{J_C} \left[A_j(\tau) \vec{\Phi}_{\pm}(v_j)^+ B_j(\tau) \vec{\Phi}_{\mp}(v_j) \right],$$

where functions $A_j(\tau)$ and $B_j(\tau)$ are found after integration of the infinite-medium Green function with the right-hand side of the SDE given above.

The general solution of the SDE can be found as a sum of a particular solution and the general solution of a homogeneous equation:

$$\vec{S}_{\pm}(\tau) = \vec{S}_{\pm}^h(\tau) + \vec{S}_{\pm}^p(\tau),$$

which includes $8N$ unknown constants $A_j, B_j, j = 1, 2, \ldots, 4N$.

Step 5. To find constants A_j, B_j, boundary conditions must be applied. In particular, one should take into consideration that there is no diffused light coming to the top of a scattering layer and the diffused light from underneath a scattering layer is just due to the light reflection from the Lambertian surface with the spherical albedo A. Boundary conditions in the notation as for step 3 are (Siewert, 2000): $\vec{S}_k^m(0,\mu) = \vec{O}$,

$$\vec{S}_k^m(\tau_0, -\mu_\sigma) = 2A\delta_{0m}\delta_{1k}\hat{I}\left\langle \mu_0\hat{D}\vec{F} \exp(-\tau_0/\mu_0) + \sum_{p=1}^{N} w_p\mu_p\vec{S}_k^m(\tau_0,\mu_p) \right\rangle,$$

where $\vec{O}$ is the zero vector and $\hat{I} = \mathrm{diag}\{1, 0, 0, 0\}$, $\sigma = 1, 2, \ldots, N$. To find the required constants from appropriate system, one can use the subroutines DGETRF and DGETRS from the LAPACK package (Anderson et al., 1995).

Step 6. To find solution not only for Gauss angles μ_σ but for all possible angles, a post-processing procedure as described by Siewert (2000) can be used.

The steps 1–6 outlined above are realized in the code SCIAPOL_1.0 , which is freely available at www.iup.physik.uni-bremen.de/~alexk. The code is capable to find the Stokes vector at any point $\vec{r}$ and at any direction specified by angles (ϑ, φ) inside a light scattering medium and also at its upper and lower boundaries for the case of a homogeneous isotropic symmetric turbid medium illuminated by the monodirectional unpolarized wide light beam. The accuracy of the code was checked against benchmark results of Siewert (2000). Five first digits coincide with corresponding results given by Siewert (2000). This confirms a high accuracy of the DOT implementation in SCIAPOL_1.0.

3.7.2 The accuracy of the scalar approximation

The VRTE enables the solution of important problems of aerosol optics in particular and atmospheric optics in general. For instance, one can study the accuracy of the scalar radiative transfer equation with respect to the calculation of the diffuse light field intensity $I(\vartheta, \phi)$. Clearly, the accuracy will depend on the phase matrix. Let us consider the accuracy of the scalar approximation for the calculation of the diffuse light intensity for a number of typical phase matrices relevant to solar light propagation in the terrestrial atmosphere. We start from the phase matrix corresponding to the molecular light scattering. Generally, the phase matrix of an unit volume of the atmospheric air can be represented as the weighted sum of the molecular and aerosol scattering contribution. The weighting procedure is identical to that described for aerosol mixtures in Chapter 2. For the moment, we will neglect the aerosol contribution and consider the accuracy of the scalar approximation for molecular scattering assuming $\omega_0 = 1$. The problem of polarized radiative transfer in a molecular scattering atmosphere is the most extensively studied among all other vector transport problems. Both analytical results (Chandrasekhar, 1950; Sobolev, 1956; van de Hulst, 1980) and extensive tables are available (Coulson et al., 1960) for both intensity and polarization characteristics of light reflected and transmitted by a molecular atmosphere. It should be noted that remote sensing of aerosols and clouds from space requires a subtraction of the molecular scattering signal from the total measured radiance. This is done using so-called pre-calculated look-up tables (LUTs) of the Rayleigh intensity depending on the ground elevation, solar and viewing angles, the relative azimuths between the Sun and receiver positions, and sensing wavelengths (Hsu et al., 2004). Mishchenko and Travis (1997) clearly showed that the VRTE (and not the SRTE) should be used in the construction of LUTs. This is also confirmed by Fig. 3.2(a). This figure shows the dependence of the relative error of the scalar approximation ε in percent as the function of the optical thickness of the molecular atmosphere at the nadir observation conditions, several solar angles and $\omega_0 = 1$. The value of the error in percent is defined as: $\varepsilon = 100(1 - I_s/I_v)$, where I_s is the intensity obtained from the solution of the scalar problem and I_v is the intensity of the reflected light obtained using the VRTE. We see that the error ε could be quite large especially for the nadir illumination and grazing incidence angles at $\vartheta = 0°$. The error has a maximum around the optical thickness 1, which roughly corresponds to the wavelength 320 nm in the case of the terrestrial atmosphere (Bucholtz, 1995). Note that wavelengths $\lambda < 320$ nm are not extensively used for lower atmosphere remote sensing due to the interference of generally unknown in advance ozone absorption. Generally, the error of the scalar approximation can be neglected for $\tau < 0.02$, which corresponds to the wavelengths $\lambda \geq 800$ nm. So the scalar approximation can be used in the construction of LUTs in the near-infrared. However, this is not the case in the visible or, especially, in the UV region of the electromagnetic spectrum. This fact is still often ignored in modern aerosol retrieval techniques, which leads to the increased errors in retrievals depending on the wavelength used and the illumination/observation conditions of the scene under study.

It is known that the intensity of singly scattered light does not differ in either scalar or vector formulations under unpolarized light illumination conditions. Hence, we have: $\varepsilon \to 0$ as $\tau \to 0$. The intensity of reflected light for semi-infinite media only weakly depends on the vector nature of light fields. This is due to the randomization of light polarization states by multiple light scattering. Therefore, one could expect the existence of the maxima of the absolute error somewhere in the transition zone from single scattering regime to highly developed multiple light scattering. Such maxima are clearly visible in our calculations presented in Fig. 3.2 at $0.2 \leq \tau \leq 1$.

It is interesting to see that the scalar approximation can either overestimate the reflected light intensity (e.g., for the solar zenith angles 0–45 degrees and the nadir observation) or underestimate the reflected light intensity (e.g., for solar zenith angles larger than 45 degrees and the nadir observation; see Fig. 3.2). To understand this feature better, we plotted the value of ε as the function of the scattering angle θ in Fig. 3.3. The scattering angle is defined as: $\theta = \arccos(-\mu\mu_0 + ss_0 \cos\phi)$, where $s = \sqrt{1 - \mu^2}$, $s_0 = \sqrt{1 - \mu_0^2}$. In particular, it follows at $\mu = 1$ as used in Fig. 3.3 that $\theta = \pi - \vartheta_0$. It follows from Fig. 3.3 that the error of the scalar approximation (at $\varepsilon > 0$) is largest at $\theta = \pi$. This coincides with the findings shown in Fig. 3.2 because the case of the solar zenith angle equal to zero at the nadir observation corresponds to the exact backscattering geometry (see the upper curve in Fig. 3.2). It is known that the scalar and vector theory produce the same results for single scattering under unpolarized solar light illumination conditions. However, this is not the case for the secondary scattering and generally for the multiple light scattering regime. The

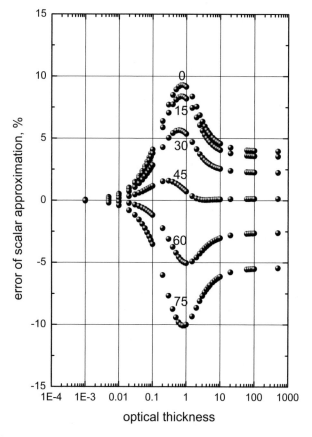

Fig. 3.2. Error of the scalar approximation at the nadir observation for several solar zenith angles as the function of the optical thickness of a scattering layer.

largest differences should occur at scattering angles close to 90 degrees because then the polarization of scattering light is almost complete for molecular scattering.

We have approximately for the upper-left 2 * 2 sub-matrix of the general Stokes matrix for molecular scattering at the scattering angle 90 degrees (Kokhanovsky, 2003):

$$\hat{C} = \begin{pmatrix} 1 & 1 \\ 1 & 1 \end{pmatrix},$$

where we neglect the constant multiplier. This matrix produces the following reduced Stokes vector $\vec{s}$ of the singly scattered light at $\theta = 90°$ for the incident unpolarized light:

$$\vec{s} = \begin{pmatrix} 1 \\ 1 \end{pmatrix}.$$

We define the reduced vector $\vec{s}$ as the Stokes vector $\vec{S}$ with neglected components U, V. Clearly, it follows for the double scattering in the same plane as for the incident light:

$$\vec{s} = \begin{pmatrix} 2 \\ 2 \end{pmatrix}.$$

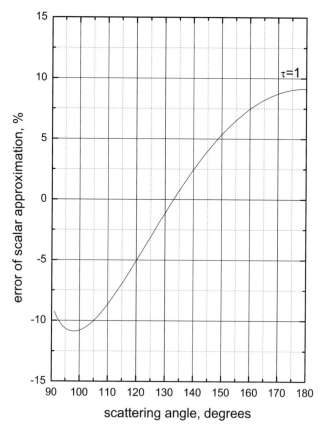

Fig. 3.3. Error of the scalar approximation at the nadir observation at the optical thickness of a scattering layer equal to 1.0 as the function of the scattering angle for the nadir observation conditions.

We have, however, for the double scalar scattering (both scatterings are at right angles and in the same scattering plane):

$$\vec{s} = \begin{pmatrix} 1 \\ 0 \end{pmatrix}.$$

So the intensity of scattered light is two times smaller for the secondary scattering in the scalar approximation as compared to the case where the vector character of scattering is fully accounted for. This explains the large positive errors in the intensity of reflected light for the backscattering geometry as shown in Fig. 3.3 ($I_s < I_v$). This physical insight is due to Mishchenko et al. (1994), who also explained the minimum for the negative ε (see Fig. 3.3(a)) using similar arguments as given above for two scatterings at right angles to each other but with the rotation of the scattering plane by 90 degrees for the second scattering. This gives, for the intensity of scattered light, $I_v = 0$. The normalized intensity I_s is still equal to 1 for the scalar case. This explains the overestimation of the reflected light intensity at $\vartheta_0 = 75°$ by the SRTE in Fig. 3.2. Then the scattering angle is 105°, which is close to 90°. Note that the minimum in Fig. 3.3 occurs at $\theta = 97°$ and not at exactly 90° due to the influence of triple and higher-order scatterings.

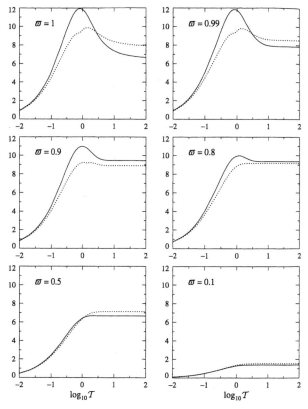

Fig. 3.4. Maximum overestimation (in percent, solid curves) and maximum underestimation (in percent, dotted curves) errors of the scalar approximation versus optical thickness for molecular atmosphere at different values of the single scattering albedo and the black underlying surface (Mishchenko et al., 2006).

We have studied the dependence of the error ε for the nadir observation conditions $(\vartheta = 0°)$. It is of importance to understand what happens with the error for arbitrary values of the angles $(\vartheta_0, \vartheta, \phi)$. For this, 3-D plots $\varepsilon(\vartheta_0, \vartheta, \phi)$ are necessary. However, it is also instructive to look in the maximal values of the error at given τ_0 and arbitrary angles $(\vartheta_0, \vartheta, \phi)$. This is illustrated in Fig. 3.4, where the solid curve gives the maximal value ε_+ of the function $\varepsilon(\vartheta_0, \vartheta, \phi)$ for different values of τ_0 and ω_0. This error can be called the maximum underestimation error. One concludes that the error of scalar approximation decreases with the absorption in the scattering layer. The error ε_+ is positive, which underlines the fact that for any τ_0, ω_0 at varying viewing and observation conditions, one always find the case when $\varepsilon(\vartheta_0, \vartheta, \phi) > 0$ and, therefore, the scalar approximation underestimates the solution of the vector radiative transfer equation. For the cases considered in Fig. 3.2(a), the error ε_+ will coincide with the upper curve. This underlines the fact that it does not give the maximal error of the approximation, which is given, for example, by the lower curve in Fig. 3.2(a) at $\tau_0 \geq 1$. Therefore, it is also instructive to introduce the maximal overestimation error ε_-, which will correspond to the lower curve in Fig. 3.2(a). This error defined as $\varepsilon_- = -\max\{\varepsilon_\phi\}$ for any pair (ϑ_0, ϑ) is presented by the dotted line in Fig. 3.4. It corresponds to the lower curve in Fig. 3.2 taken with the opposite sign. Here ε_ϕ

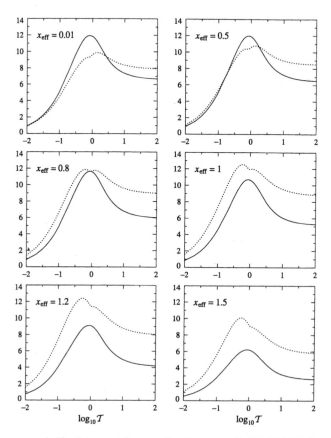

Fig. 3.5(a). The same as in Fig. 3.4 except for aerosol media at $x_{\text{ef}} = [0.01, 1.5]$ (Mishchenko et al., 2006).

is the minimal value of $\varepsilon(\vartheta_0, \vartheta, \phi)$ for a given (ϑ_0, ϑ) and the azimuth varying in the range $[0, \pi]$. For instance, one can conclude from Fig. 3.4 that the error ε is bounded by values -10 and $+12$ percent at $\tau_0 = 1, \omega_0 = 1$. Adding the Lambertian surface underneath the scattering layer leads to the decrease of the error of the scalar approximation (Mishchenko et al., 2006). This is due to the further randomization of light scattering processes.

Results similar to those shown in Fig. 3.4 but for aerosol media are presented in Figs. 3.5(a) and (b). Calculations were performed assuming water aerosol with the refractive index equal to 1.33 and the lognormal distribution with the effective variance equal to 0.1 and different values of the effective size parameter $x_{\mathrm{ef}} = \pi d_{\mathrm{ef}} / \lambda$, where $d_{\mathrm{ef}} = 2 a_{\mathrm{ef}}$. These calculations suggest that the scalar equation can be successfully used for particles with an effective diameter $d_{\mathrm{ef}} \geq \lambda$, which is certainly the case for water clouds in the visible and near-infrared and also for oceanic and dust-type aerosols. However, for the fine-mode aerosol, $d_{\mathrm{ef}} = 0.2 \ \mu\mathrm{m}$, and the error increases. Fine-mode aerosols never exist in isolation in the atmosphere. Therefore, the validity of the scalar approximation depends on the fraction of the coarse mode. For the cases of aerosol media with average diameters smaller than the wavelength of the incident light one must consider the VRTE for the calculation of diffuse light intensity. Otherwise, errors can be on the order of 10 % depending on the optical thickness and also on incidence and observation angles.

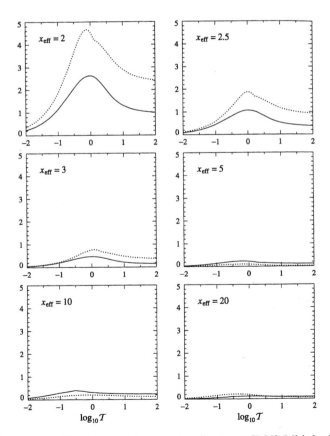

Fig. 3.5(b). The same as in Fig. 3.4 except for aerosol media at $x_{\mathrm{ef}} \in [2,20]$ (Mishchenko et al., 2006).

3.7.3 The accuracy of the single scattering approximation

The optical thickness of molecular atmosphere for wavelengths usually used for the tropospheric aerosol remote sensing ($\lambda < 320$ nm) is smaller than 1.0. It is smaller than 0.1 for
wavelengths larger than 700 nm (see Appendix). So it is of importance to see if the single
scattering approximation (SSA) can be used to find the Stokes vector of scattered light for
molecular atmosphere in this case. Let us consider the accuracy of the SSA for the normalized Stokes vector of the reflected light $\vec{S}_r$ at the illumination of a scattering medium by an
unpolarized light beam and for the nadir observation conditions. The corresponding approximate equation can be written in the following form for the Stokes vector of the reflected light in the framework of the SSA (Hansen and Travis, 1974):

$$\vec{S}_r = \frac{\omega_0 \hat{P}(\mu, \mu_0, \varphi)}{4(\mu + \mu_0)} \{1 - \exp(-m\tau)\} \vec{F},$$

$m = (\mu^{-1} + \mu_0^{-1})$ and

$$\vec{F} = \pi I_0 \vec{J}, \quad \vec{J} = \begin{pmatrix} 1 \\ 0 \\ 0 \\ 0 \end{pmatrix},$$

for unpolarized light illumination conditions.

The components of the normalized Stokes vector of the reflected light $\vec{S}_r^*$ are defined as
follows:

$$S_{r1}^* = \pi I_r / F_0 \mu_0, \quad S_{r2}^* = \pi Q_r / F_0 \mu_0, \quad S_{r3}^* = \pi U_r / F_0 \mu_0, \quad S_{r4}^* = \pi V_r / F_0 \mu_0,$$

where the Stokes vector of reflected light is given as

$$\vec{S}_r = \begin{pmatrix} I_r \\ Q_r \\ U_r \\ V_r \end{pmatrix}.$$

We present results of calculations using the SSA for the reflection function $R \equiv S_{r1}^*$ and the
polarization difference $D = -S_{r2}^*$ in Fig. 3.6 at $\omega_0 = 1$. The phase matrix of the Rayleigh
scattering in the form (Kokhanovsky, 2003):

$$\hat{P}(\theta) = \frac{3}{4} \begin{pmatrix} 1 + \cos^2\theta & -\sin^2\theta & 0 & 0 \\ -\sin^2\theta & 1 + \cos^2\theta & 0 & 0 \\ 0 & 0 & 2\cos\theta & 0 \\ 0 & 0 & 0 & 2\cos\theta \end{pmatrix}$$

was used in approximate calculations. Clearly, it follows that $S_{r3}^* = S_{r4}^* = 0$ in the case
considered. Note that we neglect here possible depolarization effects, which exist due
to the molecular anisotropy (see Appendix).

We see that the SSA underestimates both R and D and can be used with the accuracy
better than 5 % only at $\tau \leq 0.05$ (or for wavelengths larger than approximately 650 nm

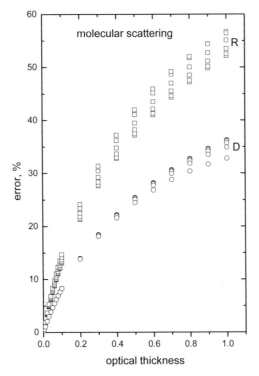

Fig. 3.6. Dependence of the error of the single scattering approximation on the optical thickness in the case of pure molecular scattering. Squares correspond to errors in the reflection function and circles give the error in the polarization difference at the nadir observation conditions. The same symbols for the fixed value of the molecular optical thickness correspond to different solar zenith angles.

(Bucholtz, 1995)) . So most of the visible and UV parts of the electromagnetic spectrum are not covered by this approximation for molecular scattering in the terrestrial atmosphere. However, Fig. 3.6 shows that the dependence of the error $\Delta = 100(1 - R_{SSA}/R_v)$ on the solar angle is not very pronounced. So we can introduce a correction multiplier $f(\tau)$ to the SSA at the average angle $45°$. Then it follows that

$$R = R_{ss}f(\tau), \quad D = S_{ss}\psi(\tau),$$

where we found using the parameterization of numerical results that

$$f(\tau) = 1 + \frac{(7 - \tau)\tau}{5}, \quad \psi(\tau) = 1 + a\tau - b\tau^2 + c\tau^3$$

and $a = 0.864$, $b = 0.442$, $c = 0.133$. The accuracy of the modified SSA approximation at $\vartheta = 0°$ is given in Figs. 3.7(a) and (b) both for R and D. We see that the accuracy is much improved in comparison with the standard SSA. The error is smaller than 5% for most solar illumination conditions and $\tau \leq 1$, which corresponds approximately to the wavelengths $\lambda \geq 320$ nm in the case of the terrestrial atmosphere (Bucholtz, 1995). Similar results can be obtained for other observation conditions as well. Also they can be improved, if necessary, to cover still larger values of τ.

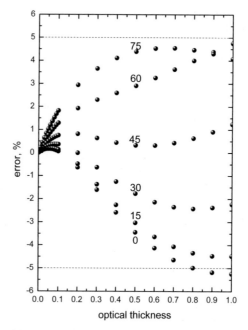

Fig. 3.7(a). Dependence of the error of the modified single scattering approximation for the value of the reflection function on the optical thickness in the case of pure molecular scattering at the nadir observation conditions and the solar zenith angles equal to 0, 15, 30, 45, 60, and 75 degrees.

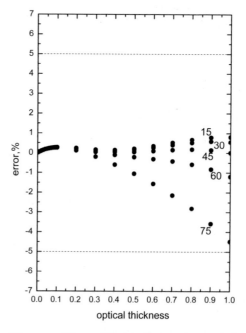

Fig. 3.7(b). Dependence of the error of the modified single scattering approximation for the polarization difference on the optical thickness in the case of pure molecular scattering at the nadir observation conditions and the solar zenith angles equal to 0, 15, 30, 45, 60, and 75 degrees.

3.7.4 The intensity and degree of polarization of light reflected from an aerosol layer

The VRTE can be used for studies of angular characteristics of light reflected and transmitted by an aerosol layer. This is illustrated in Figs. 3.8(a) and (b) for the reflection function and the degree of polarization of the reflected light for two distinct types of aerosol media with refractive indices $1.43 - i\chi$ and $1.53 - i\chi$, where $\chi = 0.006$. It was assumed that particles are polydispersed and the particle size distribution is given by

$$f(a) = Aa^{v} \exp\left(-v\frac{a}{a_0}\right)$$

with $A = v^{v+1}a_0^{-v-1}\Gamma^{-1}(v+1)$, $v = 6$, $a_0 = 400$ nm. The calculations are performed using Mie theory (van de Hulst, 1957) at the wavelength 550 nm. It follows that both R and D differ considerably depending on the value of the refractive index of particles. It means that the value of n can be retrieved from reflected light measurements.

It follows from Fig. 3.8(a) that the reflectances are lower at the refractive index $n = 1.43$ as compared to the case $n = 1.53$. Therefore, if the refractive index 1.53 is assumed in the construction of corresponding LUTs of aerosol retrieval algorithms for particles with lower n (e.g., due to uptake of water), then the retrieved aerosol optical thickness will be underestimated. The error will be even larger if the degree of polarization

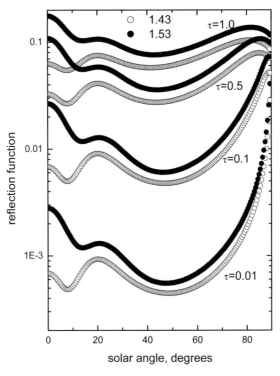

Fig. 3.8(a). Dependence of the aerosol reflection function on the solar zenith angle for values of AOT equal to 0.001, 0.1, 0.5, and 1.0, assuming nonabsorbing aerosols with refractive indices 1.43 and 1.53.

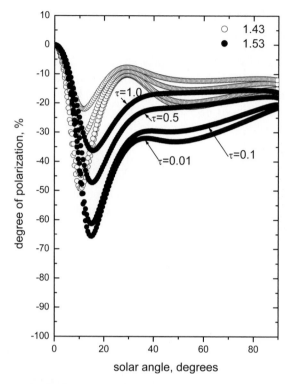

Fig. 3.8(b). The same as in Fig. 3.8(a) except for the degree of polarization of scattered light, assuming that the incident light is unpolarized.

measurements are used assuming the wrong value of the refractive index (see Fig. 3.8(b)). It also means that the refractive index of particles should be retrieved simultaneously with the AOT to avoid such biases. Corresponding retrieval algorithms have already been developed (Zhao et al., 1997). Considerable efforts should be put into their further development and application to satellite data. For this, however, one needs data from optical instruments capable of measuring both polarization and intensity of reflected light (Deschamps et al., 1994). The possibility of changing the viewing conditions for the same ground target (see, for example, Moroney et al., 2002, and references therein) is also of a great importance for this task. The general behavior of the degree of polarization curves $P(\vartheta_0)$ is quite different for different refractive indices (Zhao et al., 1997). This can also be used to find n and constrain corresponding LUTs.

An interesting feature of the degree of polarization shown in Fig. 3.8(b) is that it is negative for almost all solar angles. This is due to the corresponding behaviour of the degree of polarization for singly scattered light in the case studied (see Fig. 3.9), which is negative or partially linearly polarized in the direction parallel to the meridional plane for all scattering angles. This is in contrast with molecular and cloud scattering (see Fig. 3.9). We see that the measurements of sign and angular dependence of the degree of polarization is of a great importance for atmospheric remote sensing. The polarization characteristics of aerosol media vary considerably depending on the chemical composition of aerosol particles, their morphology, the shape of particles, and their size (Junge, 1963).

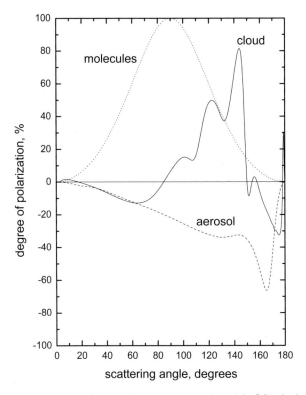

Fig. 3.9. Dependence of the degree of polarization on the scattering angle θ for single light scattering of unpolarized radiation by molecules, water cloud droplets, and the aerosol medium calculated using Mie theory at the wavelength 550 nm. Water droplets have the effective radius 6 μm and the refractive index 1.333. Aerosol particles have the effective radius 0.6 μm and the refractive index 1.53–0.006i. Particle size distributions are given by the gamma distribution with the half-width parameter υ equal to 6.0 both for clouds and aerosols.

Chapter 4. Fourier optics of aerosol media

4.1 Main definitions

The results presented in the previous chapter were aimed at the description of angular, spectral, and polarization characteristics of multiply scattered light beams both inside and at the boundaries of an aerosol layer. They are quite general and can be used for the solution of numerous practical problems ranging from the optimization of a turbid layer with respect to the characteristics of reflected and transmitted light beams to the problems of vision in atmosphere. The focus of this chapter is on the image transfer theory.

Let us imagine that there is a distant object observed through a thick layer of haze. Depending on the microstructure of the haze and also on its optical thickness, one can see details of the object with a higher or lower contrast with respect to the background. Let us imagine that we observe a bar chart shown in Fig. 4.1 through an aerosol layer. Clearly, the increase of the aerosol optical thickness will lead to the loss of contrast defined as $\kappa = (E_{max} - E_{min})/(E_{max} + E_{min})$, where E_{min} is the minimal value of the brightness and E_{max} is its maximal value in the object plane. This is due to the transversal diffusion of photons from bright areas in Fig. 4.1 to black areas (as seen by a distant observer) during light propagation from the object to the eye through an aerosol layer. The effect is demon-

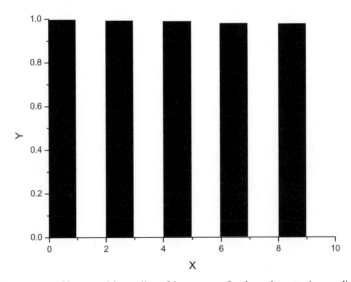

Fig. 4.1. The sequence of bars used in studies of image transfer through scattering media.

strated in Fig. 4.2, where two photographs were taken at the same location but on a clear and on a hazy day. Clearly, at some critical value of the optical thickness of a scattering layer, an eye is no longer capable of distinguishing an object against the background. Then only a uniform distribution is seen and one cannot recognize that there is actually an object behind the scattering layer (at least with the naked eye). The disappearance of the bar chart

a)

b)

Fig. 4.2. The view of distant objects seen through a heavy aerosol layer (a) and also for a clear day (b).

shown in Fig. 4.1 as seen by a distant object will depend not only on the aerosol optical thickness and its microstructure but also on the distance $x = l/2$ between two black bars. It is obvious that objects with smaller values of the period l will disappear more quickly with the increase of the aerosol optical thickness as compared to the case with larger values of l. This is due to the fact that the transversal diffusion of photons washes out the fine details of the object more quickly then the large-scale details. This statement can be formulated also in terms of the spatial frequency $v = 1/l$ measured in mm^{-1}. Therefore, a scattering layer can be considered as a filter of high spatial frequency. Signals with low spatial frequencies will pass through a layer with little disturbance but high frequencies will be attenuated considerably.

Therefore, a scattering aerosol layer can be characterized not only by its transmittance t but also by the modulation transfer function (MTF) $T(v)$, which describes the efficiency of transfer of a signal of a given spatial frequency through an aerosol layer. This function can be introduced in the following way.

Any diffused source of light can be considered as a superposition of point light sources. Thus, in linear optical systems the image of such an object (e.g., a wall illuminated by the Sun) with the irradiance $a_0(\vec{r}')$ is a linear superposition of images of point sources. This can be represented as

$$a(\vec{r}) = \int\limits_{-\infty}^{\infty} \int\limits_{-\infty}^{\infty} S(\vec{r}, \vec{r}') a_0(\vec{r}') \, d\vec{r}', \qquad (4.1)$$

where the point spread function (PSF) $S(\vec{r}, \vec{r}')$ describes the process of the transformation of the object irradiance $a_0(\vec{r}')$ in the initial object plane to the image irradiance $a(\vec{r})$ in the image plane. The point spread function is a main notion of the image transfer theory (ITT) (Zege et al., 1991). Let us assume that the PSF depends only on the difference $\vec{r} - \vec{r}'$. This means that the image of the point-source object changes only in location, not in functional form, as the point source explores the object field. Therefore, we can write Eq. (4.1) in the following form:

$$a(\vec{r}) = \int\limits_{-\infty}^{\infty} \int\limits_{-\infty}^{\infty} S(\vec{r} - \vec{r}') a_0(\vec{r}') d\vec{r}' \qquad . \qquad (4.2)$$

Let us introduce the Fourier transforms of the correspondent functions:

$$a(\vec{v}) = \int\limits_{-\infty}^{\infty} \int\limits_{-\infty}^{\infty} a(\vec{r}') e^{-i2\pi\vec{v}\,\vec{r}'} \, d\vec{r}', \qquad (4.3)$$

$$a_0(\vec{v}) = \int\limits_{-\infty}^{\infty} \int\limits_{-\infty}^{\infty} a_0(\vec{r}') e^{-i2\pi\vec{v}\,\vec{r}'} \, d\vec{r}', \qquad (4.4)$$

$$S(\vec{v}) = \int\limits_{-\infty}^{\infty} \int\limits_{-\infty}^{\infty} S(\vec{r}') e^{-i2\pi\vec{v}\vec{r}'} \, d\vec{r}'. \qquad (4.5)$$

For the bar chart shown in Fig. 4.1, one can use just one spatial frequency v_x along the axis OX. There is no changes in the brightness in the direction of the axis OY. In reality, however, the brightness can change in the direction of the axis OY and, therefore, we need to introduce the spatial frequency v_y for the description of this process and the vector $\vec{v}$ for the arbitrary distribution of the brightness in the image plane.

It follows from Eq. (4.2) after its multiplication by $e^{-i2\pi\vec{v}\,\vec{r}}$ from both sides and integration with respect to $\vec{r}$:

$$a(\vec{v}) = \int\limits_{-\infty}^{\infty} \int\limits_{-\infty}^{\infty} d\vec{r} \int\limits_{-\infty}^{\infty} \int\limits_{-\infty}^{\infty} d\vec{r}' S(\vec{r} - \vec{r}') a_0(\vec{r}') e^{-i2\pi\vec{v}\vec{r}}. \tag{4.6}$$

Let us introduce a new variable $\vec{p} = \vec{r} - \vec{r}'$ instead of $\vec{r}$. Then it follows:

$$a(\vec{v}) = \int\limits_{-\infty}^{\infty} \int\limits_{-\infty}^{\infty} d\vec{p} \int\limits_{-\infty}^{\infty} \int\limits_{-\infty}^{\infty} d\vec{r}' S(\vec{p}) a_0(\vec{r}') e^{-i2\pi(\vec{v}\vec{r}' + \vec{v}\vec{p})} \tag{4.7}$$

or

$$a(\vec{v}) = S(\vec{v}) a_0(\vec{v}). \tag{4.8}$$

Therefore, the integration procedure as shown in Eq. (4.2) is substituted by the multiplication operation in Fourier space. This makes all calculations much simpler. The function $S(\vec{v})$ is called the optical transfer function (OTF). This is a central notion of Fourier optics of aerosol media. In particular, it can be proven that for the cases of several aerosol layers and also to account for the OTF of other atmospheric processes like turbulence and the OTF of imaging instruments, one must use the simple product of all relevant OTFs to have the total OTF of the whole propagation channel from the object to the image space.

The knowledge of the OTF of a given aerosol layer enables an immediate calculation of the PSF using the inverse Fourier transform of Eq. (4.5):

$$S(\vec{p}) = \int\limits_{-\infty}^{\infty} \int\limits_{-\infty}^{\infty} S(\vec{v}) e^{i2\pi\vec{v}\vec{p}} d\vec{v}, \tag{4.9}$$

which differs just by sign of the exponent appearing in the integrand from the Fourier transform defined by Eq. (4.5). For the simplicity of notation here and also later in the text, we use the same symbol for the function and its Fourier transform (e.g., S).

One concludes that the determination of the irradiance distribution in the image plane for an arbitrary distribution of the irradiance $a_0(\vec{r}')$ in the object plane is reduced just to a simple integration as shown in Eq. (4.2). All physics of the problem including the dependence of the image properties on the size, shape, and chemical composition of aerosol particles in a medium between an object and a receiver is contained in just one function $S(\vec{v})$. This explains the background behind extensive studies of OTFs for different scattering media including clouds, aerosols, and ocean (Zege et al., 1991). The OTF is a complex function, which can be presented in the following form:

$$S(\vec{v}) = a(\vec{v}) \exp[i\varphi(\vec{v})]. \tag{4.10}$$

The modulus $a(\vec{v})$ normalized to its value at zero frequency is known as the modulation transfer function (MTF):

$$T(\vec{v}) = \frac{a(\vec{v})}{a(0)}. \tag{4.11}$$

The function $\varphi(\vec{v})$ is called the phase transfer function (PTF). The MTF describes the change of contrast with the spatial frequency due to light scattering, diffraction and propagation effects.

Let us assume that the PSF is a symmetric function with respect to the azimuth ϕ. In this case PSF can be described just by the modulus $r \equiv |\vec{r}|$. Also the OTF will depend only on $v \equiv |\vec{v}|$ and has a circular symmetry with respect to the azimuth ψ in the image plane. Therefore, introducing polar coordinates $(r \cos \phi, r \sin \phi)$ and $(v \cos \psi, v \sin \xi)$ in the object and image planes, respectively, one derives:

$$a(v) = \int_0^{2\pi} d\phi \int_0^\infty dr \, ra(r) \exp[-i2\pi v r \cos(\psi - \phi)] \tag{4.12}$$

and taking into account the integral representation of the Bessel function

$$J_0(x) = \frac{1}{2\pi} \int_0^{2\pi} \exp(-ix \cos(\psi - \phi)) \, d\phi, \tag{4.13}$$

it follows for the azimuthally symmetric OTF that

$$a(v) = 2\pi \int_0^\infty J_0(vr) a(r) r \, dr. \tag{4.14}$$

Clearly, the OTF is a real function and the PTF is equal to zero in this case. It also means that the MTF coincides with the normalized OTF for the case under consideration. Symmetric PSFs occur in many atmospheric optics applications.

In a similar way one can show that

$$a(r) = 2\pi \int_0^\infty J_0(vr) a(v) v \, dv. \tag{4.15}$$

So the distribution of irradiance in the image plane can be presented as a single integral (Fourier–Bessel transform) of the OTF for the case under consideration. There is no difference between the transform and inverse-transform operations for circularly symmetric functions. Denoting the Fourier–Bessel transform of the function $a(r)$ as $B(a(r))$, one easily derives:

$$B(a(nr)) = n^{-2} a\left(\frac{v}{n}\right). \tag{4.16}$$

This means that a narrowing of the point spread function n times will lead to the broadening of the OTF n times and *vice versa*. Clearly, narrower PSFs will enable a better image transfer from an object to an image plane. In particular, it follows from Eq. (4.2) that, if the

PSF can be substituted by the delta function $\delta(\vec{r} - \vec{r}')$, the object will be seen in the image plane without any disturbance $(a(r) \equiv a_0(r))$. Therefore, if an aerosol layer has a broad optical transfer function, this will lead to better imaging of the fine details of an object through this layer. The important problem is, therefore, to study the influence of sizes of particles, their refractive indices and also aerosol optical thickness on the OTF and its half-width. This problem is addressed in the next section assuming that particles in an aerosol layer are characterized by phase functions highly extended in the forward direction, which is the case, for example, for coarse aerosols.

4.2 Image transfer through aerosol media with large particles

4.2.1 Theory

Let us assume that the scattering layer is illuminated along normal by a light source with the intensity distribution $I_0(\vec{r}, \vec{s})$. Unlike most of the problems considered above, we need to account for the horizontal inhomogeneity of multiply scattered light field. This inhomogeneity arises solely due to boundary conditions. Instead of uniform illumination of the aerosol upper boundary (e.g., by solar light) , we will consider the case of an arbitrary illumination by the light field $I_0(\vec{r}, \vec{s})$ dependent both on the position $\vec{r}$ and the direction $\vec{s}$ (e.g., for incident laser beam).

The main equation describing the problem can be written in the following form:

$$\left(\vec{s}\vec{\nabla}\right)I(\vec{r}, \vec{s}) + k_{\text{ext}}I(\vec{r}, \vec{s}) - \frac{k_{\text{sca}}}{4\pi} \int_{4\pi} I(\vec{r}, \vec{s}')p(\vec{s}', \vec{s}) \, d\Omega' = 0, \qquad (4.17)$$

where we neglect the vector nature of light for the simplification of derivations. The task is to find the angular distribution of the light field $I(\vec{r}, \vec{s})$ at a given point $\vec{r}$ in the direction specified by the vector $\vec{s}$. An important approximate solution can be derived, if one is interested in the distribution $I(\vec{r}, \vec{s})$ close to the axis of the incident narrow light beam for vectors $\vec{s}$ directed along OZ or in the directions almost parallel to OZ. We present vectors $\vec{r}$ and $\vec{s}$ as

$$\vec{r} = x\vec{e}_x + y\vec{e}_y + z\vec{e}_z, \qquad (4.18)$$

$$\vec{s} = s_x\vec{e}_x + s_y\vec{e}_y + s_z\vec{e}_z, \qquad (4.19)$$

where $(\vec{e}_x, \vec{e}_y, \vec{e}_z)$ are unity vectors directed along the axes OX, OY, OZ. Axes OX and OY specify the plane perpendicular to OZ. We can write in the spherical coordinate system:

$$s_x = \sin\theta\cos\varphi, \quad s_y = \sin\theta\sin\varphi, \quad s_z = \cos\theta. \qquad (4.20)$$

Here φ is the azimuthal angle and θ is the angle between the axis OZ and the observation direction. We will assume that $\theta \to 0$ in our derivations. Hence, the corresponding approximation is called the small-angle approximation. Let us introduce the vector

$$\vec{\nabla}_\perp \equiv \vec{e}_x\frac{\partial}{\partial x} + \vec{e}_y\frac{\partial}{\partial y}. \qquad (4.21)$$

Then it follows that

$$\left(\vec{s}\vec{\nabla}_\perp\right)I(z,\vec{p},\vec{s}) + \frac{\partial I(z,\vec{p},\vec{s})}{\partial z} + k_{ext}I(z,\vec{p},\vec{s}) - \frac{k_{sca}}{4\pi}\int\limits_{-\infty}^{\infty}ds_x\int\limits_{-\infty}^{\infty}ds_y I(z,\vec{p},\vec{s}')p(\vec{s}'-\vec{s}) = 0,$$

(4.22)

where we used the fact that $ds_x\,ds_y = \cos\theta\sin\theta\,d\theta\,d\varphi = \cos\theta\,d\Omega \approx d\Omega$ and $s_z \approx 1$ as $\theta \to 0$. Also we assumed that the phase function depends only on the difference vector $\vec{d} = \vec{s} - \vec{s}'$ and introduces the transverse vector $\vec{p} = x\vec{e}_x + y\vec{e}_y$. We use infinite limits of integration because the contribution of photons located at large distances from the axis OZ is low. Clearly, our assumptions are valid only if light scattering occurs predominantly in the forward direction and this is really the case for coarse aerosols. The approximation considered is not valid in deep layers of a scattering medium (e.g., at $\tau \geq 5$) because then light deviates from the axis OZ considerably.

Instead solving Eq. (4.22) and then performing the Fourier transform to derive the OTF, we will first apply the Fourier transform to Eq. (4.22) and then solve the corresponding equation in the Fourier space. This will enable us to derive the analytical equation for the double Fourier transform of light intensity $\tilde{I}(z,\vec{v},\vec{q})$ defined as

$$\tilde{I}(z,\vec{v},\vec{q}) = \int\limits_{-\infty}^{\infty}\int\limits_{-\infty}^{\infty}d\vec{s}\int\limits_{-\infty}^{\infty}\int\limits_{-\infty}^{\infty}d\vec{p}\,I(z,\vec{p},\vec{s})\,e^{-i2\pi(\vec{v}\vec{p}+\vec{q}\vec{s})}.$$

(4.23)

Because the irradiance E is determined as

$$E = \int\limits_{-\infty}^{\infty}\int\limits_{-\infty}^{\infty}d\vec{s}I(z,\vec{p},\vec{s}),$$

(4.24)

it follows for OTF that

$$S(\vec{v}) = I(z,\vec{v},0).$$

(4.25)

Therefore, to calculate the OTF, one needs to find $\tilde{I}(z,\vec{v},\vec{q})$ and substitute $\vec{q}$ by zero. Applying the Fourier transform with respect to $\vec{p}$ to Eq. (4.22), we have, using definitions specified in Table 4.1:

$$\hat{\Lambda}I(z,\vec{v},\vec{s}) - \frac{k_{sca}}{4\pi}\int\limits_{-\infty}^{\infty}\int\limits_{-\infty}^{\infty}d\vec{s}'I(z,\vec{v},\vec{s}')p(\vec{s}-\vec{s}') = 0,$$

(4.26)

where $\hat{\Lambda} \equiv (\partial/\partial z) + \sigma_{ext} - i\vec{s}\vec{v}$. This equation can be simplified using the substitution:

$$I(z,\vec{v},\vec{s}) = D(z,\vec{v},\vec{s})\exp(i2\pi z\vec{q}\vec{s} - \tau),$$

(4.27)

where $\tau = \sigma_{ext}z$. Thus, one obtains:

$$\frac{dD(z,\vec{v},\vec{s})}{dz} - \frac{k_{sca}}{4\pi}\int\limits_{-\infty}^{\infty}\int\limits_{-\infty}^{\infty}D(z,v,\vec{s}')G(z,\vec{v},\vec{s}-\vec{s}')\,d\vec{s}' = 0,$$

(4.28)

Table 4.1. Fourier transforms

No.	Fourier transform	Definition
1	$\tilde{I}(z,\vec{v},\vec{s})$	$\displaystyle\int_{-\infty}^{\infty}\int_{-\infty}^{\infty} I(z,\vec{p},\vec{s})\, e^{i\vec{v}\vec{p}}\, d\vec{p}$
2	$I(z,\vec{p},\vec{s})$	$\displaystyle\frac{1}{4\pi^2}\int_{-\infty}^{\infty}\int_{-\infty}^{\infty} \tilde{I}(z,\vec{v},\vec{s})\, e^{-i\vec{v}\vec{p}}\, d\vec{s}$
3	$-i\vec{s}\vec{v}\tilde{I}(z,\vec{v},\vec{s})$	$\displaystyle\int_{-\infty}^{\infty}\int_{-\infty}^{\infty} \left(\vec{s}\vec{\nabla}_{\perp}\right)I(z,\vec{p},\vec{s})\, e^{i\vec{v}\vec{p}}\, d\vec{p}$
4	$\tilde{p}(\vec{q})$	$\displaystyle\int_{-\infty}^{\infty}\int_{-\infty}^{\infty} p(\vec{s})\, e^{i\vec{q}\vec{s}}\, d\vec{s}$
5	$\tilde{h}(\vec{v}) = \tilde{f}(\vec{v})\tilde{g}(\vec{v})$	$\displaystyle h(\vec{\beta}) = \int_{-\infty}^{\infty}\int_{-\infty}^{\infty} f(\vec{a})g(\vec{\beta}-\vec{a})\, d\vec{a}$
6	$\tilde{\tilde{G}}(z,\vec{v},\vec{q})$	$\displaystyle\int_{-\infty}^{\infty}\int_{-\infty}^{\infty} p(\vec{s})e^{i(\vec{q}-z\vec{v})\vec{s}}\, d\vec{s}$
7	$\tilde{D}(z,\vec{v},\vec{s})$	$\displaystyle\frac{1}{4\pi^2}\int_{-\infty}^{\infty}\int_{-\infty}^{\infty} \tilde{\tilde{D}}(z,\vec{v},\vec{q})\, e^{-i\vec{q}\vec{s}}\, d\vec{q}$
8	1	$\displaystyle\int_{-\infty}^{\infty}\int_{-\infty}^{\infty} \delta(\vec{s})e^{-i\vec{v}\vec{s}}\, d\vec{s}$

where

$$G(\vec{s}-\vec{s}') = p(\vec{s}-\vec{s}')\exp(i\vec{v}z(\vec{s}-\vec{s}')). \qquad (4.29)$$

Let us apply the Fourier transform with respect to $\vec{s}$ to the just-derived equation. The integral in Eq. (4.28) can be transformed using the convolution property 5 in Table 4.1. Then it follows that

$$\frac{d\tilde{D}(z,\vec{v},\vec{q})}{dz} - \frac{k_{\text{sca}}}{4\pi}\tilde{D}(z,\vec{v},\vec{q})\tilde{G}(z,\vec{v},\vec{q}) = 0. \qquad (4.30)$$

The tilde above a symbol means the double Fourier transform operation (both with respect to $\vec{p}$ and $\vec{s}$) of the corresponding function similar to that shown in Eq. (4.23). In particular, it follows that

$$\tilde{D}(z,\vec{v},\vec{q}) = \int_{-\infty}^{\infty}\int_{-\infty}^{\infty} D(z,\vec{v},\vec{s})\, e^{-i2\pi\vec{q}\vec{s}}\, d\vec{s} \qquad (4.31)$$

and, therefore,

$$D(z,\vec{v},\vec{s}) = \int_{-\infty}^{\infty}\int_{-\infty}^{\infty} \tilde{D}(z,\vec{v},\vec{q})\, e^{i2\pi\vec{q}\vec{s}}\, d\vec{q} \quad . \qquad (4.32)$$

The expression for $\tilde{G}$ is given in Table 4.1. Comparing lines 3 and 5 in Table 4.1, we derive: $\tilde{G} \equiv p(\vec{q} - \vec{v}z)$. Therefore, it follows that

$$\tilde{D}(z,\vec{v},\vec{q}) = \tilde{D}(0,\vec{v},\vec{q}) \exp\left\{ \frac{k_{sca}}{4\pi} \int_0^z p(\vec{q} - \vec{v}z)\, dz \right\} \tag{4.33}$$

and also

$$\tilde{D}(z,\vec{v},0) = \tilde{D}(0,\vec{v},0) \exp\left\{ \frac{k_{sca}}{4\pi} \int_0^z p(\vec{v}z)\, dz \right\}. \tag{4.34}$$

This solves the problem at hand. Indeed, the value of $D(z,\vec{v},\vec{s})$ in Eq. (4.27) can be found using the inverse Fourier transform

$$D(z,\vec{v},\vec{s}) = \int_{-\infty}^{\infty} \int_{-\infty}^{\infty} \tilde{D}(z,\vec{v},\vec{q})\, e^{i2\pi\vec{q}\vec{s}}. \tag{4.35}$$

Therefore, it follows that

$$I(z,\vec{v},\vec{s}) = D(z,\vec{v},\vec{s}) \exp(i2\pi z\vec{v}\vec{s} - \tau)$$

Assuming that $I(0,\vec{p},\vec{s}) = \delta(\vec{p})\delta(\vec{s})$ and using property 8 in Table 4.1, we derive:

$$\tilde{I}(z,\vec{v},\vec{q}) = \exp\left\{ -k_{sca}z + \frac{k_{sca}}{4\pi} \int_0^z \tilde{p}(\vec{q} - \vec{v}(z'-z))\, dz' \right\}. \tag{4.36}$$

Calculations of $\tilde{p}$ can be simplified assuming the circular symmetry of the phase function: $p(\vec{s}) \equiv p(|\vec{s}_\perp|) = p(s)$. Then one obtains:

$$p(\kappa) = 2\pi \int_0^{\infty} p(s)J_0(\kappa s)s\, ds. \tag{4.37}$$

Therefore, we can write:

$$\tilde{I}(z,\vec{v},\vec{q}) = \exp\left\{ -k_{sca}z + \frac{k_{sca}}{2} \int_0^z dz' \int_0^{\infty} d\theta p(\theta)J_0((q - vz')\theta)\theta \right\}. \tag{4.38}$$

It follows from this equation at $\vec{v} = \vec{q} = \vec{0}$ that

$$\tilde{I}\left(z,\vec{0},\vec{0}\right) = \exp\left\{ -k_{sca}z + \frac{k_{sca}}{2} \int_0^z dz' \int_0^{\infty} d\theta p(\theta)\theta \right\} \tag{4.39}$$

or $\tilde{I}\left(z,\vec{0},\vec{0}\right) = 1$, where we accounted for the phase function normalization condition:

$$\frac{1}{2} \int_0^{\infty} p(\theta)\theta\, d\theta = 1. \tag{4.40}$$

The small-angle approximation for the OTF follows from Eq. (4.38) at $\vec{q} = 0$. Namely, one derives:

$$S(z, v) = \exp\left\{-k_{sca}z + \frac{k_{sca}}{2}\int_0^z dz' \int_0^\infty d\theta p(\theta)J_0(vz'\theta)\theta\right\}. \tag{4.41}$$

or

$$S(z, v) = \exp\{-\tau[1 - \omega_0 B(v, z)]\}, \tag{4.42}$$

where

$$B(v, z) = \frac{1}{z}\int_0^z dz' p(v(z - z')). \tag{4.43}$$

Let us introduce a new variable $\varsigma = 1 - z'/z$. Then it follows that

$$B(v, z) = \int_0^1 d\varsigma p(vz\varsigma). \tag{4.44}$$

We see that $B(v, z)$ depends on the dimensionless frequency $\omega = vz$. The same is true for the OTF. Therefore, it follows that

$$S(\omega) = \exp\{-\tau[1 - \omega_0 B(\omega)]\}, \tag{4.45}$$

where

$$B(\omega) = \int_0^1 p(\omega\varsigma)\, d\varsigma \tag{4.46}$$

or

$$B(\omega) = \frac{1}{2}\int_0^1 d\varsigma \int_0^\infty d\theta p(\theta)J_0(\omega\varsigma\theta)\theta \tag{4.47}$$

Eq. (4.45) makes it possible to derive the irradiance in the image plane (see Eqs (4.2), (4.9)) if one knows the irradiance in the object plane. Also it follows that

$$a_0(\omega) = S^{-1}(\omega)a(\omega). \tag{4.48}$$

Therefore, the initial image can be reconstructed, if the OTF is known.

It follows at $\omega = 0$: $S(0) = \exp(-k_{abs}z)$. Therefore, only absorption processes are responsible for the OTF reduction at zero frequency. This is the consequence of the approximation used, which does not account for backscattering effects.

Let us approximate the phase function of a cloud medium as

$$p(\theta) = 4a^2 \exp(-a^2\theta^2). \tag{4.49}$$

Then it follows that

$$p(\varsigma\omega) = \exp\left\{-\frac{\varsigma^2\omega^2}{4a^2}\right\} \tag{4.50}$$

and, therefore:

$$B(\kappa) = \frac{a\sqrt{\pi}}{\kappa} \operatorname{erf}\left[\frac{\omega}{2a}\right], \tag{4.51}$$

where the error function

$$\operatorname{erf}(u) = \frac{2}{\sqrt{\pi}} \int_0^u \exp(-\phi^2)\, d\phi \tag{4.52}$$

is introduced. So we obtain the following analytical expression for the OTF:

$$S(\kappa) = \exp\left\{ -\tau\left[1 - \frac{a\omega_0\sqrt{\pi}}{\kappa} \operatorname{erf}\left(\frac{\omega}{2a}\right)\right]\right\}. \tag{4.53}$$

In particular, it follows as $\kappa \to 0$ that

$$\operatorname{erf}\left[\frac{\omega}{2a}\right] \approx \frac{\omega}{a\sqrt{\pi}}\left[1 - \frac{\omega^2}{12a^2}\right] \tag{4.54}$$

and, therefore,

$$S(\omega) = \exp\left\{ -\tau\left[1 - \omega_0\left(1 - \frac{\omega^2}{12a^2}\right)\right]\right\}. \tag{4.55}$$

It follows for nonabsorbing media that $S(\omega) = \exp\{-\Xi\omega^2\}$, where $\Xi = \tau/12a^2$. We see that the distribution $S(\kappa)$ has the Gaussian shape at small dimensionless frequencies ω. Larger particles in aerosol media are characterized by more extended phase functions. Therefore, a must be larger for larger particles (see Eq. (4.49)). This also means that the OTF is larger for larger particles. This will lead to a better image quality for media having larger particles.

The Gaussian shape of the OTF as $\omega \to 0$ is characteristic for the small-angle approximation in general. The result does not depend on the assumption on the phase function. It can be demonstrated in the following way. It is known that the Bessel function can be written in series with respect to its argument as follows:

$$J_0(x) = \sum_{n=0}^{\infty} a_n x^{2n}, \tag{4.56}$$

where

$$a_0 = 1, \quad a_1 = -\frac{1}{4}, \quad a_2 = \frac{1}{4^2(2!)^2}, \quad a_3 = -\frac{1}{4^3(3!)^2}, \quad \text{etc.}$$

The substitution of Eq. (4.56) enables to evaluate the double integral in Eq. (4.47) analytically. The answer is

$$B(\omega) = \sum_{n=0}^{\infty} \frac{a_n \omega^{2n}}{2n+1} \langle \theta_{2n} \rangle, \tag{4.57}$$

where

$$\langle \theta_{2n} \rangle = \frac{1}{2} \int_0^\infty \theta^{2n+1} p(\theta) \, d\theta. \tag{4.58}$$

The moments $\langle \theta_{2n} \rangle$ can be easily found if the phase function is known. In particular it follows that $\langle \theta_0 \rangle = 1$ and

$$\langle \theta_2 \rangle = \frac{1}{2} \int_0^\infty \theta^3 p(\theta) \, d\theta. \tag{4.59}$$

The value of $\langle \theta_2 \rangle$ can be related to the average cosine of the scattering angle g for phase functions highly peaked in the forward direction. Indeed, it follows that

$$g = \frac{1}{2} \int_0^\infty \cos \theta p(\theta) \theta \, d\theta. \tag{4.60}$$

Using the approximation valid at small scattering angles: $\cos \theta \approx 1 - (\theta^2/2)$, one derives:

$$g = 1 - \frac{\langle \theta_2 \rangle}{2}. \tag{4.61}$$

This means that $\theta_2 \approx 2(1 - g)$ for phase functions highly extended in the forward direction.

Therefore, it follows at small spatial frequencies that

$$B(\omega) = 1 - \frac{\omega^2 \langle \theta_2 \rangle}{12} \tag{4.62}$$

and, therefore,

$$S(\omega) = \exp\left\{ -\tau_{abs} - \frac{\omega^2 \tau_{sca}}{12} \langle \theta_2 \rangle \right\}, \tag{4.63}$$

where $\tau_{abs} = k_{abs} z$, $\tau_{sca} = k_{sca} z$. So the Gaussian shape of $S(\omega)$ at small spatial frequencies is a general feature of small scattering approximation irrelevant to a particular angular behavior of a given highly extended in the forward direction phase function. Also it follows for the MTF as $\omega \to 0$ in the framework of SAA that

$$T(\omega) = \exp\left\{ -\frac{\omega^2 \tau_{sca}}{12} \langle \theta_2 \rangle \right\}. \tag{4.64}$$

The real aerosol phase functions can be modeled by the approximation given in Eq. (4.49) in theoretical studies of image transfer through aerosol media. Clearly, the elongation of phase functions in the forward direction depends on the size of scatterers. The analysis of Mie computations shows that $1 - p(\theta)/p(0)$ is proportional to $(ka_{ef})^{-2}$ at small scattering angle and, therefore, a is the inversely proportional to the square root of the effective radius of particles. Clearly, Eq. (4.49) does not allow us to study effects of the complex refractive index on the image transfer. Then exact Mie computations of the phase functions or moments (4.58) are needed.

4.2.2 Geometrical optics approximation

The general behavior of the OTF with respect to the complex refractive index of particles and also their sizes can be obtained using the geometrical optics approximation valid for particles having sizes much larger than the wavelength of the incident light (e.g., image transfer in fogs). Then the angular scattering coefficient

$$k_{\text{sca}}(\theta) = \frac{4\pi N I(\theta)}{k^2} \tag{4.65}$$

in the forward direction can be represented by the sum of three components, namely, diffracted, transmitted, and reflected light components. Here, N is the number of particles in a unit volume and $k = 2\pi/\lambda$, λ is the wavelength. It follows (van de Hulst, 1957, Kokhanovsky, 2006) that the function $I(\theta)$ for monodispersed spherical particles much larger than the wavelength (the radius $a \gg \lambda$) can be represented as the sum of three components in the small-angle region:

$$I = I^{\text{d}} + I^{\text{t}} + I^{\text{r}}, \tag{4.66}$$

where the interference between different components is neglected and

$$I^{\text{d}}(\theta) = \frac{x^2 J_1^2(\theta x)}{\theta^2}, \tag{4.67}$$

$$I^{\text{r}}(\theta) = \frac{x^2}{4} \sum_{j=1}^{2} \left[\frac{N_j^2 (1-q^2)^{1/2} - (n^2 - q^2)^{1/2}}{N_j^2 (1-q^2)^{1/2} + (n^2 - q^2)^{1/2}} \right]^2, \tag{4.68}$$

$$I^{\text{t}}(\theta) = \left(\frac{2n}{n^2 - 1} \right)^4 \frac{(nq-1)^3 (n-q)^3 (1+q^4) x^2}{8q^5 (1+n^2 - 2nq)^2} \exp(-hc(x)), \tag{4.69}$$

where $x = 2\pi a/\lambda$, $q = \cos(\theta/2)$, $h = (n-q)(1+n^2 - 2nq)^{-1/2}$, $N_1 = 1$, $N_2 = n^2$, $c = 4\chi x$ and it is assumed that $\chi \ll n$.

Now we note that the expression for the OTF can be written in the following form:

$$S(\omega) = \exp\{(\kappa(\omega) - k_{\text{ext}})L\}, \tag{4.70}$$

where

$$\kappa(\omega) = \frac{1}{2} \int_0^1 d\varsigma \int_0^\infty d\theta k_{\text{sca}}(\theta) J_0(\omega\varsigma\theta)\theta. \tag{4.71}$$

It follows that the function $\kappa(\omega)$ can be presented as the sum of three components for diffracted, refracted and reflected light:

$$\kappa(\omega) = \kappa^{\text{d}}(\omega) + \kappa^{\text{t}}(\omega) + \kappa^{\text{r}}(\omega), \tag{4.72}$$

where

$$\kappa_{\text{d,t,r}}(\omega) = \frac{2\pi N}{k^2} \int_0^1 d\varsigma \int_0^\infty d\theta I_{\text{d,t,r}}(\theta) J_0(\omega\varsigma\theta)\theta. \tag{4.73}$$

The integral for the diffraction component can be found analytically using the fact that

$$2 \int_0^\theta J_1^2(\theta x) J_0(\theta b)\theta \, d\theta = \left(\arccos\left(\frac{b}{2x}\right) - \frac{b}{2x}\sqrt{1 - \left(\frac{b}{2x}\right)^2}\right) u\left(\frac{b}{2x}\right), \qquad (4.74)$$

where $u(b/2x)$ equals to zero at $b \geq 2x$ and one, otherwise. Then it follows after simple derivations:

$$K_d(\omega) = \pi N a^2 q_d(\omega), \quad q_d(\omega) = \frac{4}{3\pi b} + \left(\frac{2}{\pi}\arccos b - \frac{2(2+b^2)\sqrt{1-b^2}}{3\pi b}\right) u(b), \quad (4.75)$$

where $b = \omega/2x$. Corresponding integrals for the transmitted and reflected component cannot be evaluated analytically in the general case. However, the approximate calculation is possible. The analysis of correspondent functions show that they can be substituted by exponents as follows (Zege and Kokhanovsky, 1994):

$$I^r(\theta) = \frac{x^2}{4}\exp(-a\theta), \qquad (4.76)$$

$$I^t(\theta) = Ax^2 \exp\left(-c - \beta\theta^2\right)/4, \qquad (4.77)$$

where

$$A = \frac{16n^4}{(n+1)^4(n-1)^2}. \qquad (4.78)$$

The values of a and β are tabulated by Zege and Kokhanovsky (1994). In particular they found that $a = \beta = 2.4$ at $n = 1.53$. Then one derives:

$$K_r(\omega) = \pi N a^2 q_r(\omega), K_t(\omega) = \pi N a^2 q_t(\omega), \qquad (4.79)$$

where

$$q_r(\omega) = \frac{1}{2a^2\sqrt{1 + \frac{\omega^2}{a^2}}}, \quad q_t(\omega) = \frac{A\exp(-c)}{4\omega}\sqrt{\frac{\pi}{\beta}}\, \mathrm{erf}\left(\frac{\omega}{2\sqrt{\beta}}\right) \qquad (4.80)$$

and

$$\mathrm{erf}(s) = \frac{2}{\sqrt{\pi}}\int_0^s e^{-t^2/2}\, dt \qquad (4.81)$$

is the error function. This function has the following behavior as $s \to 0$:

$$\mathrm{erf}(s) = \frac{2s}{\sqrt{\pi}}. \qquad (4.82)$$

This means that it follows at $\omega = 0$ that

$$q_t(0) = \frac{A\exp(-c)}{4\beta}. \qquad (4.83)$$

Also one easily derives:

$$q_d(0) = 1, q_r(0) = \frac{1}{2a^2}.$$ (4.84)

This means that

$$S(0) = \exp(-\tau(1 - \varpi)),$$ (4.85)

where

$$\varpi = \frac{1}{2} + \frac{1}{4a^2} + \frac{A\,e^{-c}}{8\beta}$$ (4.86)

and we accounted for the fact that it follows for the optical thickness of aerosols layers with large scatterers:

$$\tau = 2N\pi a^2.$$ (4.87)

Note that $S(0)$ coincides with the diffuse transmission coefficient of a scattering layer under normal illumination conditions (Zege et al., 1991; Zege and Kokhanovsky, 1994). We can derive using Eq. (4.70) and also formulae given above:

$$S(\omega) = \exp\left\{-\tau\left(1 - \frac{1}{2}(q_d(\omega) + q_r(\omega) + q_t(\omega))\right)\right\}.$$ (4.88)

Therefore, it follows for the modulation transfer function that

$$T(\omega) = \exp\left\{-\tau\left(\varpi - \frac{1}{2}(q_d(\omega) + q_r(\omega) + q_t(\omega))\right)\right\}.$$ (4.89)

The results of calculations of the OTF at $\tau = 1$ using Eq. (4.87) are shown in Fig. 4.3 for size parameters of 100, 500, and 1000 and a refractive index equal to 1.53. The following approximation of the error function was used:

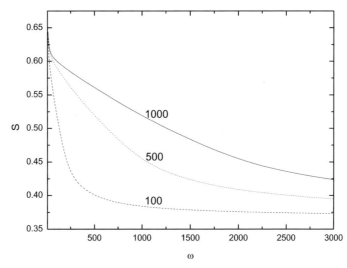

Fig. 4.3. The optical transfer function for size parameters equal to 100, 500, and 1000. It is assumed that n = 1.53 and $\tau = 1$.

$$\mathrm{erf}(s) = 1 - \left(a_1\varepsilon + a_2\varepsilon^2 + a_3\varepsilon^3\right)\exp(-s^2), \qquad (4.90)$$

where $\varepsilon = (1 + ws)^{-1}, w = 0.470\,47, a_1 = 0.348\,024\,2, a_2 = -0.095\,879\,8, a_3 = 0.747\,855\,6$. It follows that for larger particles values of the OTF are larger. This means that aerosol media with larger particles make smaller distortions of images of distant objects. This coincides with earlier findings reported above. Clearly, it follows that as $\omega \to \infty$: $S \to \exp(-\tau)$ independently of the size of particles (at a given optical thickness). Then the OTF is determined mostly by the attenuated direct light beam.

Chapter 5. Optical remote sensing of atmospheric aerosol

5.1 Ground-based remote sensing of aerosols

5.1.1 Spectral attenuation of solar light

Aerosol optical and microphysical characteristics can be deduced from the analysis of solar light scattered or attenuated by the atmosphere using instruments placed on the ground, a ship, an aircraft or a satellite. In this section we will consider the ground-based passive techniques. Active remote sensing techniques based on the analysis not solar light (like it is the case for passive techniques) but on the study of lidar signals transmitted or reflected by aerosol media are considered in the next section.

The most common passive technique involves the measurement of the transmitted direct solar light beam. It is known that the solar beam intensity I attenuates in the atmosphere exponentially:

$$I = I_0 \exp(-\tau/\cos \vartheta_0),$$

where ϑ_0 is the solar zenith angle and I_0 is the top-of-atmosphere (TOA) irradiance. It follows from this equation for the atmospheric optical thickness:

$$\tau = \cos \vartheta_0 \ln(I_0/I)$$

The value of I_0 can be estimated from measurements of I at several solar zenith angles. Then it follows that

$$\ln I = \ln I_0 - M\tau,$$

where M is the air mass factor equal to $1/\cos \vartheta_0$. It is supposed that the value of τ does not change during measurements and, therefore, the plot of $\ln I$ as the function of M (Langley plot) will enable the determination of $\ln I_0$ by the extrapolation of the measurements to the case $M = 0$. Therefore, measurements at low solar elevation angles (counted from the horizon) are needed to find the TOA irradiance. For correct measurements especially at high latitudes the corrections to the value of M due to the sphericity of atmosphere and refraction effects must be taken into account. The measured value of τ contains also contributions from gaseous absorbers. Therefore, usually one selects channels less effected by gaseous absorption to perform measurements. Then correspondent correction algorithms are applied to derive the aerosol optical thickness as the difference of measured optical thickness with that due to Rayleigh scattering (see Appendix Table A1) and gaseous absorbers like ozone, nitrous oxide, and water vapor, always present in the atmosphere in varying quantities.

The instruments measuring the spectral atmospheric transmittance are called sun photometers. Sun photometers are commercially available. For instance the network of Sun photometers called AERONET (Holben et al., 1998) consists of a number (about 200) identical Sun photometers placed at different locations worldwide. AERONET provides not only spectral AOT but also derived aerosol properties such as single-scattering albedo, asymmetry parameter, phase function, and size distributions of aerosol particles at a given location. Vertically integrated quantities are given. The results are accessible in real time via website www.gsfc.aeronet.com. Such a comprehensive list of retrieved parameters is due to the fact that the spectral diffuse scattered light is also measured by a Sun photometer in the almucantar and principal plane. The principle plane contains the normal to the surface and also the direction to the sun. The almucantar measurements are performed in the following way: The Sun photometer is directed to the Sun and then the measurement is performed using the azimuthal scan of the sky without changing the observation zenith angle of the instrument. The network hardware consists of identical automatic Sun–sky scanning spectral radiometers (CIMEL Electronique 318A; see Fig. 5.1) owned by national agencies and universities. Data from this collaboration provides globally distributed, near real time observations of aerosol spectral optical depths, aerosol size distributions, and precipitable water in diverse aerosol regimes. The data undergo preliminary processing (real-time data), reprocessing, quality assurance, archiving and distribution from NASA's Goddard Space Flight Center master archive and several other data bases (see http://aeronet.gsfc.nasa.gov/). A full list of the sites where AERONET instruments are positioned is given at http://aeronet.gsfc.nasa.gov/photo_db/site_index.html (see Fig. 5.2)

The CIMEL Electronique 318A spectral radiometer is a solar-powered, weather-hardy, robotically pointed Sun and sky spectral radiometer. The radiometer makes two basic measurements, either direct Sun or sky, both within several programmed sequences. The direct Sun measurements are made in eight spectral bands requiring approximately 10 seconds. Eight interference filters at wavelengths of 340, 380, 440, 500, 670, 870, 940 and 1020 nm are located in a filter wheel which is rotated by a direct drive stepping motor. The 940-nm channel is used for column water abundance determination. A pre-programmed sequence of measurements is taken by these instruments starting at an airmass, M, of 7 in the morning and ending at an airmass of 7 in the evening. Optical thickness is calculated from the spectral extinction of direct beam radiation at each wavelength based on the Beer–Bouguer law. Attenuation due to Rayleigh scattering, absorption by ozone (from interpolated ozone climatology atlas), and gaseous pollutants (e.g., NO_2) is estimated and removed to isolate the aerosol optical thickness. A sequence of three such measurements, taken 30 seconds apart, creates a triplet observation per wavelength. During the large airmass periods direct Sun measurements are made at 0.25 airmass intervals, while at smaller airmasses the interval between measurements is typically 15 minutes. The time variation of clouds is usually greater than that of aerosols causing an observable variation in the triplets that can be used to screen clouds in many cases. Additionally, the 15-minute interval allows for a longer temporal frequency check for cloud contamination.

In addition to the direct solar irradiance measurements that are made with a field of view of 1.2 degrees (approximately twice the value of the solar disk angle as observed from ground), these instruments measure the sky radiance in four spectral bands (440, 670, 870 and 1020 nm) along the solar principal plane (i.e., at constant azimuth angle, with varied scattering angles) up to nine times a day and along the solar almucantar (i.e., at constant

Fig. 5.1. CIMEL Sun photometer performing measurements in the vicinity of the North Pole (courtesy G. Heygster).

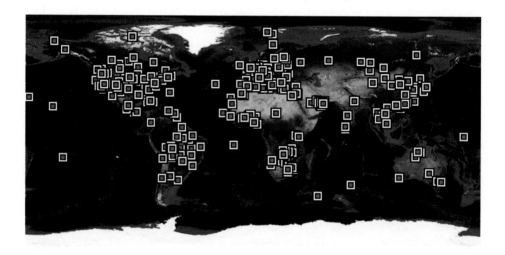

Fig. 5.2. The distribution of CIMEL Sun photometers operating in the framework of AERONET (Holben et al., 1998).

elevation angle, with varied azimuth angles) up to six times a day. The approach is to acquire aureole and sky radiance observations through a large range of scattering angles from the Sun through a constant aerosol profile to retrieve size distribution, phase function and aerosol optical thickness. For cloud-free conditions, eight almucantar sequences are made daily at an optical airmass of 4, 3, 2 and 1.7, both morning and afternoon. The details of phase function retrieval using almucantar measurements are given by Wendisch and von Hoyningen-Huene (1994) and also by von Hoyningen-Huene and Posse (1997) .

One important point of the extinction measurement is the correction for the scattered light. Clearly, some portion of the diffuse sky light will also enter an instrument leading to possible biases in the retrieved AOT. This is why the instruments are constructed in such a way that their FOV is comparable with the solar disk angular dimension. This minimizes the aureole scattering effects. Large particles such as dust are characterized by narrow scattering diagrams. Therefore, corrections of measurements must be performed to insure that the measured signal is due to extinction alone and the scattering contribution is removed. The corresponding technique for both single and multiple scattering regimes is described below.

The power as received by a ground photometer looking in the direction of the Sun can be expressed as:

$$F = \int_{\Omega_0} AI \, d\Omega, \tag{5.1}$$

where A is the receiving cross-section, $d\Omega = \sin\vartheta \, d\vartheta \, d\phi$ is the elementary solid angle, ϑ is the zenith angle, ϕ is the azimuth, and I is the light intensity in the field of view of the instrument defined by the solid angle Ω_0. It follows from Eq. (5.1) that

$$F = \Sigma \int_{\Omega_0} I \, d\Omega, \tag{5.2}$$

where the area Σ is a characteristic of a given instrument. The intensity as received by a photometer can be presented as a sum of the direct light component I_{dir} and the diffused intensity I_{dif}. One can easily derive the following expression for the direct light intensity:

$$I_{dir} = E_0 \exp(-x)\delta(\Omega_0 - \Omega), \tag{5.3}$$

where $\delta(\Omega_0 - \Omega)$ is the delta function, $x = \tau/\mu_0$, μ_0 is the cosine of the incidence angle, τ is the optical depth along the local vertical, and E_0 is the top-of-atmosphere solar irradiance. It follows for the diffused light component in the framework of the single scattering approximation assuming that the zenith observation and incidence angles coincide:

$$I_{dif} = \frac{\omega_0 E_0 p(\theta) x \exp(-x)}{4\pi}, \tag{5.4}$$

where $\omega_0 = k_{sca}/k_{ext}$ is the single-scattering albedo, k_{sca} is the scattering coefficient, k_{ext} is the extinction coefficient, $p(\theta)$ is the phase function, and θ is the scattering angle. We obtain from Eqs (5.2)–(5.4):

$$F = \Sigma E_0 \exp(-x)(1 + fx), \tag{5.5}$$

where

$$Y = \frac{\omega_0}{2} \int_0^{\theta_0} p(\theta) \sin \theta \, d\theta \qquad (5.6)$$

and θ_0 is the half FOV angle. Taking into account that $fx \to 0$ for singly scattering media, we have approximately from Eq. (5.5):

$$F = \Sigma E_0 \exp(-(1 - Y)x) \qquad (5.7)$$

and, therefore, $x = (1 - Y)^{-1}x_0$, where $x_0 = \ln(\Sigma E_0/P) \equiv \tau_0/\mu_0$ and τ_0 is the so-called apparent optical thickness. Also we can write:

$$\tau = C\tau_0, \qquad (5.8)$$

where

$$C = \frac{1}{1 - Y} \qquad (5.9)$$

is the correction factor (CF). Eqs (5.8) and (5.9) were derived by Shiobara and Asano (1994) in a way similar to that shown above. They are often used for studies of diffuse light corrections to Sun photometry and pyrheliometry.

Clearly, it follows that $Y = 0$ at $\theta_0 = 0$ (see Eq. (5.6)) and $\tau = \tau_0$ then. In reality, $\theta_0 \neq 0$, $0 < Y < 1$ and the true value of τ is larger than the apparent optical thickness τ_0. The value of C takes values between close to 1 and 2 for most practical cases, depending on the size of particles and the actual value of the FOV angle. Even larger values of C are possible, if θ_0 is not small.

Let us estimate Y for large spherical particles with the radius r much larger than the wavelength λ. Because $\theta_0 \to 0$ for modern spectrophotometers, we can use the following approximation (Shifrin, 1951) for the normalized Mie intensity $i(\theta)$ in the small-angle scattering region ($\theta \to 0$) for a spherical particle with the radius a and arbitrary complex refractive index m:

$$i(\theta) = \frac{\rho^4}{4} \Phi^2(\theta\rho), \qquad (5.10)$$

where $\rho = ka$, $k = 2\pi/\lambda$, and

$$\Phi(\theta\rho) = \frac{2J_1(\theta\rho)}{\theta\rho}, \qquad (5.11)$$

where $J_1(\theta\rho)$ is the Bessel function. Eq. (5.10) has a high accuracy as $\theta \to 0$, $\rho \to \infty$, and $p \equiv 2|m - 1|\rho \to \infty$.

The phase function of the spherical polydispersion is defined as:

$$p(\theta) = \frac{2\pi N \int_0^\infty f(a)(i_1 + i_2) \, da}{k^2 k_{\text{sca}}}, \qquad (5.12)$$

where $f(a)$ is the particle size distribution (PSD) and

$$k_{\text{sca}} = N \int_0^\infty \pi a^2 f(a) Q_{\text{sca}}(a) \, da. \tag{5.13}$$

Here Q_{sca} is the scattering efficiency factor determined from the Mie theory. The simple analytical expression for the scaled phase function $\widehat{p}(\theta) = \omega_0 p(\theta)$ can be derived at small angles θ using Eqs (5.10)–(5.12) and also the fact that $i_1 \simeq i_2 \simeq i$ at small angles. So it follows that

$$\widehat{p}(\theta) = \frac{2\langle J_1^2(k\theta r)\rangle}{\theta^2}, \tag{5.14}$$

where we used the fact that $k_{\text{ext}} = 2\pi N M_2$ (M_2 is the second moment of PSD) for large particles and angular brackets mean

$$\langle y(k\theta a)\rangle = \frac{\int_0^\infty y(k\theta a) a^2 f(a) \, da}{\int_0^\infty a^2 f(a) \, da} \tag{5.15}$$

for arbitrary function $y(k\theta a)$. One obtains from Eq. (5.14) at $\theta = 0$:

$$\widehat{p}(0) = \frac{k^2 M_{42}}{2}, \tag{5.16}$$

where $M_{42} \equiv M_4/M_2$ and

$$M_{\text{n}} = \int_0^\infty a^n f(a) \, da \tag{5.17}$$

is the n-th moment of PSD.

The accuracy of calculations according to Eq. (5.14) is demonstrated in Fig. 5.3 for the gamma PSD $f(a) = Ba^\mu \exp(-\mu a/a_0)$, where B is the normalization constant (see Table 5.1), $\mu = 6$, and a_0 is the mode radius related to the effective radius $r_{\text{ef}} \equiv M_{32}$

Table 5.1. Selected ratios of moments for gamma and lognormal PSDs (Kokhanovsky, 2007)

Parameter	Gamma PSD	Lognormal PSD
	$Ba^\mu \exp\left[-\mu\dfrac{a}{a_0}\right]$	$Ba^{-1} \exp\left[-\dfrac{1}{2\sigma^2} \ln^2\dfrac{a}{a_m}\right]$
B	$\dfrac{\mu^{\mu+1}}{a_0^{\mu+1}\Gamma(\mu+1)}$	$\dfrac{1}{\sigma\sqrt{2\pi}}$
$a_{\text{ef}} \equiv M_{32}$	$a_0\left[1+\dfrac{3}{\mu}\right]$	$a_m \exp\left[\dfrac{5}{2}\sigma^2\right]$
M_{42}	$\dfrac{\mu+4}{\mu+3}a_{\text{ef}}^2$	$a_{\text{ef}}^2 \exp\left[-\dfrac{1}{4}\sigma^2\right]$
M_{64}	$\dfrac{(\mu+5)(\mu+6)}{(\mu+3)^2}$	$r_{\text{ef}}^2 \exp\left[\dfrac{15}{4}\sigma^2\right]$

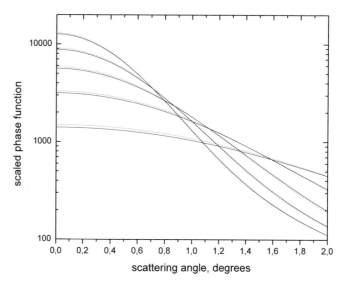

Fig. 5.3 Phase function of spherical polydispersions with the effective radius equal to 4, 6, 8, 10, and 12 μm (lower lines as $\theta \to 0$ correspond to smaller particles starting from $a_{\mathrm{ef}} = 4$ μm). Results obtained using Mie theory are shown by solid lines and the approximation is given by dotted lines. Calculations have been performed for the gamma PSD with the half-width parameter $\mu = 6$ and $\lambda = 0.5$ μm. The complex refractive index $m = 1.52–0.008i$ was used in exact numerical calculations (Kokhanovsky, 2007).

with the following analytical equation $a_{\mathrm{ef}} = a_0(1 + (3/\mu))$. The values M_{32} and M_{42} (see Eq. (5.16)) are given in Table 5.1 both for gamma and lognormal PSDs. The value of M_{62} shown in Table 5.1 appears in the asymptotic analysis of Eq. (5.14) as $\theta \to 0$. Namely it follows then that

$$\widehat{p}(\theta) = \frac{k^2 M_{42}}{2}\left(1 - \frac{k^2 M_{64}\theta^2}{4}\right). \tag{5.18}$$

We conclude from the analysis of Fig. 5.3 that Eq. (5.14) can be used instead of tedious Mie calculations at $\rho_{\mathrm{ef}} \equiv k a_{\mathrm{ef}} \gg 1$ and small scattering angles. It follows from this figure that the accuracy of Eq. (5.14) increases with the radius at small scattering angles ($\theta \to 0$) as one might expect. We also see that Eq. (5.14) works well for all angles relevant to the performance of Sun photometers with a narrow field of view. Therefore, we can use Eq. (5.14) to perform the integration as shown in Eq. (5.6). The answer is:

$$Y = \frac{1}{2}\left(1 - \langle J_0^2(k\theta_0 a)\rangle - \langle J_2^2(k\theta_0 a)\rangle\right), \tag{5.19}$$

where the meaning of angle brackets is explained above. Bessel functions J_0 and J_2 approach zero at large values of the argument $z = k\theta_0 a$. This means that it follows for very large value of z: $Y = 1/2$ and $C = 2$ (see Eq. (5.9)).

We obtain using Eqs (5.8), (5.9), and (5.19):

$$\tau = \frac{\tau_0}{1 - \frac{1}{2}\left(1 - \langle J_0^2(k\theta_0 a)\rangle - \langle J_2^2(k\theta_0 a)\rangle\right)} \tag{5.20}$$

and, therefore, finally

$$C = \frac{2}{1 + \langle J_0^2(k\theta_0 a)\rangle + \langle J_2^2(k\theta_0 a)\rangle}. \tag{5.21}$$

This gives an analytical solution for the correction factor depending on both FOV and PSD in the case of large scatterers. The results of calculations using Eq. (5.21) are shown in Fig. 5.4 both as functions of the effective radius a_{ef} (Fig. 5.4(a)) and the scaling parameter $z_{ef} = k a_{ef} \theta_0$ (Fig. 5.4(b)). Data shown in Fig. 5.2(a) have been obtained using $\theta_0 = 0.6°$, which coincides with the half-FOV angle of AERONET Cimel Sun–sky photometers.

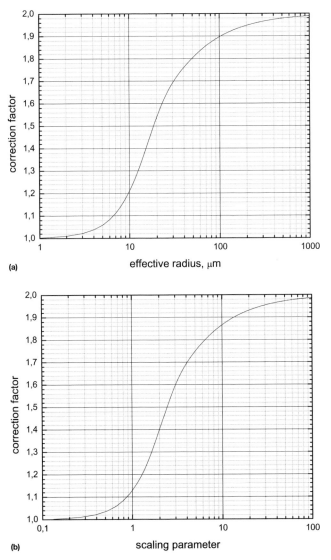

(a)

(b)

Fig. 5.4. Dependence of correction factor on effective radius ($\theta_0 = 0.6°$) (a) and scaling parameter (b). The input for calculations is the same as in Fig. 5.3 (Kokhanovsky, 2007).

One concludes from this figure that C is smaller than 1.2 for most of aerosol particles ($a_{ef} < 10$ μm). However, for the case of desert dust close to its origin the values of C can be much larger due to generally larger sizes of suspended dust grains. Yet another problem is due to thin cirrus clouds. Particles in these clouds are of the order 100–1000 μm and the correction factor is close to 2. This means that the apparent optical thickness as detected at the ground for the case of a thin cirrus cloud must be multiplied by 2. The physical analysis of the problem shows that the value of θ_0 does not effect the condition $C = 2$ as far as the whole diffraction peak is contained inside FOV of the instrument. This is the case for practically all Sun photometers. In particular, we estimate that θ_0 must be larger than 0.2° to hold the whole diffraction peak in the FOV of the instrument for the case of monodispersed spherical particles with $r = 100$ μm.

Aerosol media with large particles can also have a large optical thickness. Then the single scattering approximation considered in the previous section is not valid. Let us derive the correction factor for the case of a homogeneous plane-parallel layer with account for multiple light scattering. For this the integro-differential radiative transfer equation must be solved. This equation has the following form for the case of the nadir illumination of a scattering layer (Zege et al., 1991):

$$\mu \frac{dI(\mu, \tau)}{d\tau} = -I(\mu, \tau) + \frac{\omega_0}{2} \int_{-1}^{1} I(\tau, \eta) \bar{p}(\eta, \mu) \, d\eta, \tag{5.22}$$

where I is the total light intensity including the diffused and direct components and

$$\bar{p}(\mu, \eta) = \sum_{j=0}^{\infty} h_j P_j(\mu) P_j(\eta) \tag{5.23}$$

is the azimuthally averaged phase function, $P_j(\mu)$ is the Legendre polynomial, and h_j are coefficients in the expansion of the phase function as shown below:

$$p(\theta) = \sum_{j=0}^{\infty} h_j P_j(\cos \theta). \tag{5.24}$$

The value of μ in Eq. (5.22) is the cosine of the observation angle, which is assumed to be close to one in this work because we consider directions $\mu \approx \mu_0 = 1$ important for Sun photometry. Then it follows from Eq. (5.22)

$$\frac{dI(\mu, \tau)}{d\tau} = -I(\mu, \tau) + \frac{\omega_0}{2} \int_{-1}^{1} I(\eta, \tau) \bar{p}(\eta, \mu) \, d\eta. \tag{5.25}$$

This equation can be solved using the following substitution:

$$I(\mu, \tau) = \sum_{j=0}^{\infty} v_j(\tau) P_j(\mu) \tag{5.26}$$

and also the expansion given by Eq. (5.23). One easily derives then:

$$\frac{dv_j}{d\tau} = -v_j + \frac{\omega_0 h_j}{2j + 1} v_j, \tag{5.27}$$

where we use the normalization condition:

$$\int_{-1}^{1} P_i(\eta)P_j(\eta)\, d\eta = \frac{2}{2j+1}\delta_{ij}. \tag{5.28}$$

Here δ_{ij} is equal to one at $i = j$ and zero otherwise. It follows after integration of Eq. (5.27):

$$v_j = A_j \exp(-c_j\tau), \tag{5.29}$$

where

$$c_j = 1 - \frac{\omega_0 h_j}{2j+1} \tag{5.30}$$

and A_j are constants, which can be determined from boundary conditions. In particular, we will assume that

$$I(\mu,0) = E_0\delta(1-\mu). \tag{5.31}$$

Then using the expansion

$$\delta(1-\mu) = \frac{1}{4\pi}\sum_{j=0}^{\infty}(2j+1)P_j(\mu), \tag{5.32}$$

we derive:

$$A_j = \frac{E_0}{4\pi}(2j+1). \tag{5.33}$$

Therefore, it follows finally for the total transmitted light intensity:

$$I(\mu) = \frac{E_0}{4\pi}\sum_{j=0}^{\infty}(2j+1)\exp(-c_j\tau)P_j(\mu)\quad. \tag{5.34}$$

One obtains from Eq. (5.34) for the diffused light intensity:

$$I_{\text{dif}}(\mu) = \frac{E_0}{4\pi}\sum_{j=0}^{\infty}(2j+1)(\exp(-c_j\tau) - \exp(-\tau))P_j(\mu), \tag{5.35}$$

where we have subtracted the direct beam component (see Eqs (5.3), (5.31), and (5.32)):

$$I_{\text{dir}}(\mu) = \frac{E_0\exp(-\tau)}{4\pi}\sum_{j=0}^{\infty}(2j+1)P_j(\mu). \tag{5.36}$$

Eq. (5.35) coincides with Eq. (5.4) as $\tau \to 0$, as one can expect. That is to say, it follows in this case that

$$I_{\text{dif}}(\mu) = \frac{E_0\tau}{4\pi}\sum_{j=0}^{\infty}(2j+1)(1-c_j)P_j(\mu) \tag{5.37}$$

or after using Eqs (5.24) and (5.30):

$$I_{\mathrm{dif}}(\mu) = \frac{\omega_0 E_0 p(\theta)\tau}{4\pi}, \tag{5.38}$$

which is equivalent to Eq. (5.4) as $\tau \to 0$, $\mu_0 \to 1$. Therefore, we conclude that Eq. (5.35) has a correct limit at small optical thicknesses.

Now we take into account that we are interested in the region of very small observation angles $\vartheta = \arccos(\mu) \approx \theta$. Then the following asymptotical relationship holds: $P_j(\cos\theta) = J_0(a_j\theta)$, where $a_j = j + (1/2)$ and J_0 is the Bessel function. Therefore, Eq. (5.35) can be rewritten as:

$$I_{\mathrm{dif}}(\theta) = \frac{E_0 \exp(-\tau)}{2\pi} \sum_{j=0}^{\infty} a_j (\exp(\kappa_j\tau) - 1) J_0(a_j\theta), \tag{5.39}$$

where

$$\kappa_j = \frac{\omega_0 h_j}{2j + 1}. \tag{5.40}$$

Let us introduce the transmission function:

$$T = \frac{\pi I_{\mathrm{dif}}}{\mu_0 E_0}. \tag{5.41}$$

Then it follows that

$$T = \frac{\exp(-\tau)}{2} \sum_{j=0}^{\infty} a_j (\exp(\kappa_j\tau) - 1) J_0(a_j\theta). \tag{5.42}$$

This function does not depend on the azimuth due to the symmetry of the problem. The comparison of calculations of the transmission function according to Eq. (5.42) with exact radiative transfer numerical simulations using the vector code SCIAPOL (www.iup. physik.uni-bremen.de/~alexk) is shown in Fig. 5.5 at $\theta = 1°$. It follows that both methods

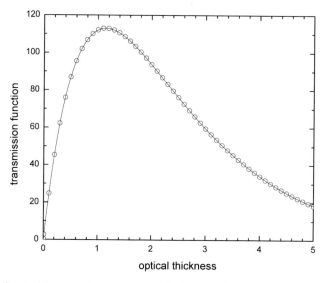

Fig. 5.5. Dependence of the transmission function at $\theta = 1°$ on optical thickness according to the approximation (line) and exact calculations (symbols) at $a_{\mathrm{ef}} = 4$ μm. Other input parameters as in Fig. 5.3 (Kokhanovsky, 2007).

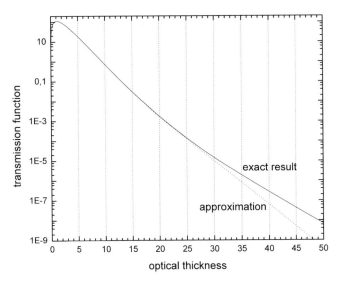

Fig. 5.6. The same as in Fig. 5.5 except in a broader range of τ (Kokhanovsky, 2007).

give the same results for all practical purposes related to Sun photometry. Therefore, Eq. (5.39) can be used instead of tedious numerical solution of the exact radiative transfer equation. It follows from Fig. 5.6 that the approximate theory described here holds up to $\tau = 25$ for the case considered above.

Summing up, we conclude that Eq. (5.39) can be used for the analysis of scattered light in the FOV of Sun photometers at any τ of practical interest. However, the angular range of applicability is generally reduced with optical thickness. This is studied in Figs 5.7–5.9. It

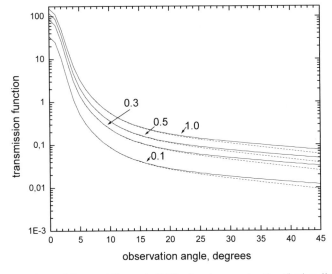

Fig. 5.7. The comparison of the exact theory (solid line) and approximation (broken line) for different values of optical thickness (0.1, 0.3, 0.5, 1.0) and $a_{ef} = 4$ μm. Other input parameters as in Fig. 5.5 (Kokhanovsky, 2007).

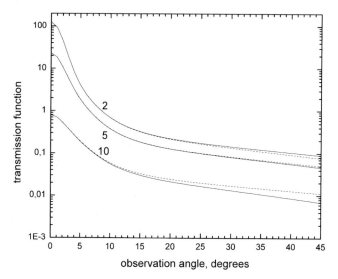

Fig. 5.8 The same as in Fig. 5.5 except at $\tau = 2, 5, 10$ (Kokhanovsky, 2007).

follows that Eq. (5.42) can be used at $\theta < 10°$ and any τ relevant to observations of the direct light with a high accuracy. As a matter of fact the numerical solutions of the radiative transfer equation can benefit from the use of Eq. (5.42) because this formula considerably increases the speed of calculations without considerable loss of accuracy at small angles.

We have proved that Eq. (5.39) is very accurate as far as its applications to Sun photometry are of concern. Therefore, it can be used to find the diffused light power F_{dif} in the FOV of the instrument (see Eq. (5.2)). It follows after the azimuthal integration that

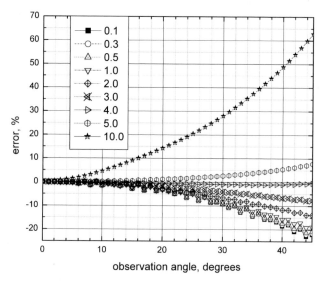

Fig. 5.9 The error of the approximation at different values of τ shown in the legend and $a_{\text{ef}} = 4$ μm (Kokhanovsky, 2007).

$$F_{\text{dif}} = \Sigma E_0 \exp(-\tau) \sum_{j=0}^{\infty} a_j \big(\exp(\kappa_j \tau) - 1\big) D_j(\theta_0), \tag{5.43}$$

where

$$D_j(\theta_0) = \int_0^{\theta_0} J_0(a_j \theta) \sin\theta \, d\theta. \tag{5.44}$$

This integral can be evaluated analytically taking into account that $\theta_0 \ll 1$. Then it follows that

$$D_j(\theta_0) = \frac{\theta_0 J_1(a_j \theta_0)}{a_j}, \tag{5.45}$$

where we used the following integral:

$$\int J_0(s) s \, ds = s J_1(s) \tag{5.46}$$

and the fact that $\sin\theta \approx \theta$ at small angles.

Finally, one obtains from Eqs (5.43) and (5.45):

$$F_{\text{dif}} = \Sigma E_0 \theta_0 \psi(\tau) \exp(-\tau), \tag{5.47}$$

where

$$\psi(\tau) = \sum_{j=0}^{\infty} \big(\exp(\kappa_j \tau) - 1\big) J_1(a_j \theta_0). \tag{5.48}$$

Therefore, the problem of the evaluation of the diffused light power as observed by a Sun photometer is reduced to the calculation of simple series. It follows from Eq. (5.47) that $F_{\text{dif}} = 0$ at $\theta_0 = 0$, the result one can expect from the general consideration of the problem at hand. Clearly, it follows for the total power:

$$F = \Sigma E_0 (1 + \theta_0 \psi(\tau)) \exp(-\tau), \tag{5.49}$$

where we added the direct light contribution. Eq. (5.49) can be also written in the following form:

$$F = \Sigma E_0 \exp(-\tau_0), \tag{5.50}$$

where

$$\tau_0 = \tau\{1 - \gamma(\tau)\} \tag{5.51}$$

and

$$\gamma(\tau) = \frac{1}{\tau} \ln(1 + \theta_0 \psi(\tau)). \tag{5.52}$$

Therefore, finally, we obtain the following expression for the correction factor $C \equiv \tau/\tau_0$ (see Eq. (5.51)):

$$C = \frac{1}{1 - \gamma(\tau)}. \tag{5.53}$$

This factor depends not only on the size of particles as in the case of the single scattering approximation (see Eq. (5.21)) but also on the value of τ. Also Eq. (5.51) is more general as compared to Eq. (5.21) because instead of approximation (5.14), coefficients h_j obtained from the Mie theory are used.

It follows from Eq. (5.48) at $\tau \ll 1$:

$$\psi(\tau) = \tau \sum_{j=0}^{\infty} \kappa_j J_1\left(a_j \theta_0\right) \tag{5.54}$$

and, therefore,

$$\gamma = \theta_0 \sum_{j=0}^{\infty} \kappa_j J_1\left(a_j \theta_0\right), \tag{5.55}$$

where we used an approximate equality $\ln(1 + x) \approx x$ valid at small x. Eq. (5.55) can be re-written in the following form (see Eq. (5.40)):

$$\gamma = \frac{\omega_0}{2} \theta_0 \sum_{j=0}^{\infty} \frac{h_j J_1\left(a_j \theta_0\right)}{a_j} \tag{5.56}$$

or taking into account Eqs (5.44), (5.45), (5.24), and (5.6), it follows: $\gamma \to f$ as $\tau \to 0$, where we also substituted the Bessel function by the Legendre polynomial, which is a valid operation at small scattering angles. Therefore, Eq. (5.9) is a particular case of Eq. (5.53) valid at small optical thicknesses. This confirms our calculations.

The dependence of C on τ (see Eqs (5.53), (5.52), and (5.48)) is shown in Fig. 5.10. Coefficients h_j (see Fig. 5.11) and also ω_0 have been calculated for the gamma PSD at $\lambda = 0.5$ μm,the dust refraction index $m = 1.52 - 0.008i$, $\mu = 6$ and several values of a_{ef}

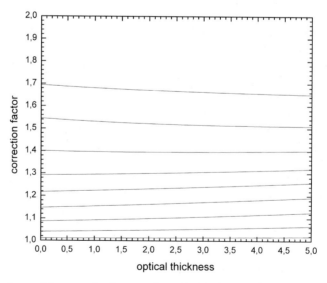

Fig. 5.10 Dependence of the correction factor on the optical thickness at $\theta_0 = 0.6$ and $a_{ef} = 2, 4, 6, 8, 10,$ 12, 15, 20, and 30 μm. The coefficients κ_j have been calculated using Mie theory for the same conditions as in Fig. 5.3. They are shown in Fig. 5.10. Lower lines correspond to smaller sizes starting from $a_{ef} = 2$ μm (Kokhanovsky, 2007).

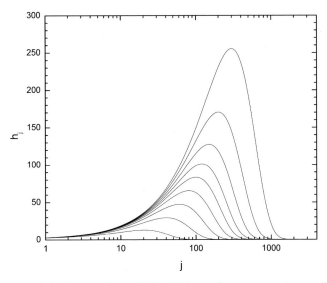

Fig. 5.11. Coefficients h_j for the cases shown in Fig. 5.9. Lower lines correspond to smaller sizes starting from $a_{ef} = 2$ μm (Kokhanovsky, 2007).

using Mie theory. The computed values of the single-scattering albedo ω_0 varied from 0.55 ($a_{ef} = 30$ μm) to 0.75 ($a_{ef} = 2$ μm).

We conclude from this figure that correction factors are not significantly affected by the value of τ. This means that simple Eq. (5.21) can be used for the estimation of correction factors in the case of aerosol media with large scatterers at arbitrary values of τ relevant to observation of attenuation of the direct light.

Results, as shown in Fig. 5.10, are difficult to obtain from numerical calculations using the exact radiative transfer equation because it involves the study of light intensity at small angles, where many hundreds of Legendre polynomials are needed to represent the small-angle peak in a correct way (see Fig. 5.11). The use of simple series given by Eq. (5.48) allows us to perform such calculations accurately and in a short time for an arbitrary number of Legendre polynomials relevant to aerosol optics problems. The shortcoming is due to the fact that only the normal illumination conditions can be analyzed using Eqs. (5.42) and (5.48). Consideration of the case of a slant illumination relevant for most cases requires numerical calculations using the exact radiative transfer equation. However, the low sensitivity of C to τ shown in Fig. 5.10 means that C is also weakly influenced by the solar zenith angle. Therefore, Eq. (5.21) can also be used for values of μ_0 different from one also at comparatively large τ. The reason for this is quite clear: the contribution of scattered light to the FOV of photometers having small value of θ_0 is mostly due to the single scattering and not to multiple light scattering.

5.1.2 Measurements of scattered light

Aerosol media not only modify the spectral composition of the direct solar beam but also they influence scattered light intensity and polarization and also they degrade image characteristics (e.g., the contrast of an object observed against a background). This means that properties of aerosol layers such as sizes of particles, and also their chemical composition and concentration, can be derived from different types of optical measurements (O'Neill and Miller, 1984; Wang and Gordon, 1993, 1994; King et al., 1999). The underlying reason for this is the fact that the average size of aerosol particles is close to the wavelength of visible light. Therefore, light characteristics are influenced by particles considerably. This is not the case in the microwave region, where the wavelength is too large for waves to be influenced by aerosols.

The determination of aerosol properties from light scattering and extinction measurements belongs to the broad class of inverse problems (Phillips, 1962; Tikhonov, 1963, Tikhonov and Arsenin, 1977; Twomey, 1963, 1977; Chahine, 1968; Turchin et al., 1970; Rodgers, 1976, 2000; Tarantola, 1987). The forward problem is aimed at calculations of light characteristics for a given ensemble of scatterers. The inverse problem is aimed at the determination of characteristics of particles from measured light extinction, scattering or polarization as functions of wavelength and corresponding angles. Clearly, the inverse problem cannot be solved in all cases, which is different from the forward problem, which always has a particular solution. Let us imagine that one measures spectral light extinction by a medium with spherical particles much larger than the wavelength for the case of thin aerosol layers. Then the extinction is determined by the average geometrical cross-section of particles independently of their particular size distribution and also refractive index. This also means that PSD and the chemical composition of particles cannot be determined from the corresponding experiment. Another difficulty is due to the fact that the dependence of the signal from the required parameter can be very weak and, therefore, if it is at a level below the experimental noise, the corresponding characteristic cannot be derived. Therefore, before actually solving the inverse problem, one must check the information content of corresponding measurements with respect to the required parameter. This can be done using either runs of the forward model for different vales of the parameter or studies of corresponding derivatives. The general strategy of solving an inverse problem is described by Twomey (1977), Tarantola (1987), and Rodgers (2000) among others.

Usually the linearization procedures are applied. For instance, let us consider the determination of the aerosol optical thickness from reflectance measurements. Clearly, the reflectance is heavily influenced by the aerosol optical thickness. Generally, it increases with the thickness of the layer. To solve this problem, the reflectance of an aerosol layer R can be presented in the Taylor series with respect to the parameter to be found (e.g., aerosol optical thickness τ_0):

$$R(\tau) = R(\tau_0) + (\tau - \tau_0)R' + \ldots,$$

where τ_0 is the assumed value of AOT, which can be taken equal to the average value for a given location (e.g., 0.2 for AOT (0.55 µm) over Europe at 0.55 µm). The derivative of the reflectance function with respect to the optical thickness R' is taken at the value of $\tau = \tau_0$. Therefore, neglecting quadratic and higher-order terms, one can derive:

$$\tau = \tau_0 + (R(\tau) - R(\tau_0))/R'.$$

This enables the calculation of AOT from the values of reflectance and its derivative with respect to the aerosol optical thickness. The calculated value of AOT is substituted in the forward problem and the deviation of the calculated reflectance from the measured one is derived. If this deviation is considerable, the next iteration is performed assuming just the derived value of AOT instead of initially assumed value τ_0. The procedure is repeated until convergence is reached. Simultaneously the errors of the inverse problem solution can be estimated as specified by Rodgers (2000). The approach described above reduces the solution of a given inverse problem to the multiple runs of forward models. The technique can be applied to the solution of a great number of inverse problems except those where nonlinear terms cannot be ignored. The formulation given above is valid for the reflectance measured at a single wavelength. One can also pose the question of the determination of a given parameter using spectral measurements. It is essential that the parameter to be retrieved does not depend on the wavelength. Take, for example, the problem of aerosol height h (e.g., dust outbreak from a desert) determination from a satellite using spectral measurements in the region where reflectance depends on h. For instance, such a dependence exists in the absorption band of oxygen. This is due to the fact that an aerosol layer screens a part of tropospheric oxygen, thereby reducing the depth of the corresponding molecular absorption line in the reflectance spectrum (e.g., around the wavelength 760 nm). Then one obtains:

$$R(h, \lambda) = R(h_0, \lambda) + (h - h_0)R'(\lambda) + \dots .$$

Measurements are taken at discrete wavelengths $\lambda_1, \lambda_2, \dots, \lambda_n$. Therefore, one can introduce the n-dimensional vector $\vec{R}$ with components equal to measured reflectances at each wavelength. Then it follows:

$$\vec{R}(h) = \vec{R}(h_0) + (h - h_0)\vec{R'} + \dots,$$

where the same notation is used for the derivative. This equation can be used for the determination of h minimizing the following cost function:

$$\Xi = \left\| \vec{\Theta} - \vec{\Psi}\zeta \right\|,$$

where $\| \ \|$ means the norm in the Euclid space of the corresponding dimension, and $\zeta = h - h_0$, $\vec{\Theta} = \vec{R} - \vec{R}(h_0)$, $\vec{\Psi} = \vec{R'}(h_0)$. The next step is to check whether the deviation of the calculated reflectance spectrum in the gaseous absorption band from the measured one is acceptable. If the standard deviation of two spectra is not small, then the next iteration must be performed until convergence is reached.

The corresponding derivatives can be calculated either using the finite different technique running the forward radiative transfer model for a given and a disturbed atmospheric state or applying the adjoint radiative transfer equation (ARTE) (Rozanov, 2006). The application of ARTE has the advantage that the speed of computation increases considerably. Although the approach based on the adjoint formulation of the radiative transfer problem is more involved from the mathematical point of view, as compared with the use of finite difference technique, which is a straightforward procedure.

The techniques described above require quite powerful computers. Yet another approach is to use so-called look-up tables. Then the reflectance is calculated for different values of the solar incidence angle, the observation angle, and the relative azimuth (e.g., for a given single-scattering albedo and the phase function). LUTs are searched until the deviation of measured reflectance and that in the LUT is no larger than a prescribed number. Also multidimensional LUTs, when several parameters are retrieved, can also be used.

The radiative transfer problems can be solved analytically in some cases. Then one can try to invert the corresponding solution analytically with respect to the required parameter. This is rarely possible; but if the problem can be solved in this way, then the solution of the inverse problem is simplified to a great extent.

A flexible inversion algorithm for the retrieval of the optical properties of atmospheric aerosol from Sun and sky radiance measurements was developed by Dubovik and King (2000). The technique is based on the previous works of King et al. (1978) and Nakajima et al. (1983, 1996). The method of King et al. (1978) is used to invert spectral aerosol optical thickness with respect to the size distribution and the method of Nakajima et al. (1983, 1996) is used to invert the angular distribution of sky radiance with account for multiple light scattering. Dubovik and King (2000) proposed an inversion technique for the simultaneous determination of the particle size distribution and the complex refractive index. The technique is currently applied to the interpretation of the AERONET Sun and sky radiance measurements (http://aeronet.gsfc.nasa.gov:8080). It is based on the analysis of the intensity of scattered light. The technique can be extended to the measurements of the degree of polarization of skylight. This will make it possible to enhance the accuracy of retrievals (e.g., with respect to the complex refractive index of particles).

The inversion technique for the determination of aerosol size distribution and the complex refractive index was also earlier proposed by Wendisch and von Hoyningen-Huene (1994). However, these authors followed a somewhat different strategy in the retrieval procedure. They first determined the aerosol particle size distribution for the assumed typical value of the aerosol complex refractive index from spectral extinction measurements. This made it possible to calculate the aerosol phase function and the single-scattering albedo, which are used together with measured aerosol optical thickness to simulate the brightness of skylight. The comparison of the measured and calculated sky brightness allows one to make a decision whether one needs to proceed to the next iteration with a new value of the refractive index or can stop iterations. In particular, the refractive index n corresponding to the minimum of the function $\delta(n)$, where δ is the root-mean-square deviation of the simulated and measured brightness of skylight, is taken as a result of inversion. The authors also elaborated an approach to deal with the case of nonspherical aerosol particles.

5.1.3 Lidar measurements

The passive measurements described above cannot be used to determine the vertical structure of aerosol. They refer to the average aerosol properties along extended vertical columns (usually from the ground to a height of several kilometers). However, in many cases information on the vertical aerosol structure is needed. This is provided by lidars. Lidar emits a monochromatic beam in the direction from which aerosol properties are required. The backscattered photons can be classified with respect to their arrival times, which also give the distance to the observation volume. This makes it possible to study the aerosol vertical or horizontal structure and its time evaluation (see, for example, http://www.awi-potsdam.de/www-pot/koldewey/tropo_archive/tropo_index.html). The receiver and emitter are usually placed almost at the same place and the scattering angle is about 180 degrees (monostatic lidars). Bistatic lidars are also used. Emitter and receiver are well-separated then.

The lidar equation can be written as (Klett, 1981):

$$P(r) = P_0 \frac{c\Delta t}{2} A \frac{\beta(r)}{r^2} \exp\left[-2 \int_0^r k_{\text{ext}}(r') \, dr'\right],$$

where $P(r)$ is the instantaneous received power at time t in the single scattering approximation, P_0 is the transmitted power at time t_0, c is the velocity of light, Δt is the pulse duration, A is the effective system receiver area, $r = c(t - t_0)/2$ is the range, $\beta(r)$ is the volume backscatter coefficient, $k_{\text{ext}}(r)$ is the extinction coefficient. An important quantity in atmospheric studies is the extinction coefficient profile $k_{\text{ext}}(r)$. This profile can be retrieved from the lidar equation assuming that $\beta(r)$ is known. Usually it is assumed that $\beta = \text{const } k_{\text{ext}}^q$, where q depends on the lidar wavelength and various properties of obscuring aerosol. Reported values of q are usually in the interval $[0.67, 1.0]$. Klett (1981) derived assuming that $q = \text{const}$:

$$k_{\text{ext}}(r) = B(r) \exp((s(r) - s(r_0))/q),$$

where

$$s(r) = \ln(r^2 P(r)),$$

$$B(r, r_0) = \left[k_{\text{ext}}^{-1}(r_0) - \frac{2}{q} \int_{r_0}^r \exp((s(r') - s(r_0))/q) \, dr'\right]$$

and $r_0 \leq r$.

This equation for the extinction profile is ill-constrained. Therefore, Klett (1981) proposed to use the modified form of this equation:

$$k_{\text{ext}}(r) = B(r, r_m) \exp((s(r) - s(r_m))/q),$$

where $r_m \geq r$ is the reference range.

Lidar measurements can be performed at several wavelengths. This enables the determination of the aerosol size distributions. The depolarization $D = \beta_\perp / \beta_\parallel$ of backscattered laser light is used to characterize the degree of nonsphericity of aerosol particles. Here $\parallel$ and $\perp$ mean the light-detection system with the polarizer oriented along the polarization of the incident light and perpendicular to it, respectively. The value of D vanishes for spherical scatterers but is considerably different from zero for nonspherical aerosols. This is illustrated in Fig. 5.12, where temporal distributions of vertical profiles of the backscattering coefficient and the depolarization ratio are given for a single location in Sawon (South Korea). The dust layer was observed over the place on November 6–8, 2005, which is confirmed by the depolarization ratio measurements both at 532 and 1064 nm. The values of D are larger for the wavelength 1064 nm as compared to the wavelength 532 nm, which is related to the diminished contributions of background aerosol composed of small spherical particles for the larger wavelength.

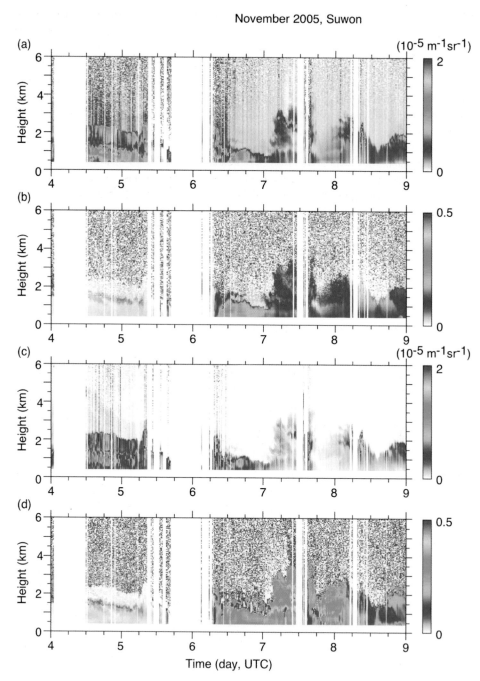

Fig. 5.12. Time–height color maps of the backscattering coefficient at β (532 nm) (a), D (532 nm) (b), β (1064 nm) (c), D (1064 nm) (d) (Sugimoto and Lee, 2006).

5.2 Satellite remote sensing of atmospheric aerosol

5.2.1 Introduction

Atmospheric aerosol forcing is one of the greatest uncertainties in our understanding of the climate system (see, for example, IPCC (2001) and also http://www.ipcc.ch/). To address this issue, many scientists are using Earth observations from satellites because the information provided is both timely and global in coverage (see, for example, Bovensmann et al., 1999; King et al., 1999; Breon et al., 2002; Kinne et al., 2006). Clearly, the interpretation of the signals detected on a satellite is much more difficult as compared to ground measurements discussed in the previous sections. Let us take the determination of the aerosol optical thickness over a snow field. The observation of the direct solar beam spectral transformation due to scattering by aerosol particles is little affected by a snow field underneath whereas the top-of-atmosphere reflectance is determined mostly by the snow properties but not the thin aerosol layer above the highly reflective surface. This underlines the problems related to the interpretation of satellite remote sensing data especially over land, where the surface contribution cannot be neglected. Therefore, the surface reflectance must be retrieved simultaneously with aerosol properties. For this advanced surface reflectance models are needed (Mishchenko et al., 1999; Kimmel and Baranoski, 2007). The matter is somewhat simpler over ocean, where the contribution from the surface in IR is very small and the signal detected on satellite is mostly due to molecular and aerosol scattering. However, this is not true for high windspeed (Koepke, 1984; Kokhanovsky, 2004c; Monahan, 2006), when foam is formed. Because the contribution of molecular atmosphere is well understood and also small, it can be subtracted from the top-of-atmosphere reflectance in IR. The aerosol reflectance obtained in such a way can be compared with pre-calculated look-up tables. This enables the determination of the aerosol optical thickness (e.g., assuming the single-scattering albedo and aerosol phase function characteristic for a given region). The results for the smaller wavelengths (e.g., in the visible) can be obtained, using the Angstrom law by extrapolation. It is desirable to derive not only the aerosol optical thickness but also the single scattering-albedo and the phase function using satellite aerosol retrieval algorithms. This is possible, however, only for spectro-photopolarimetric multiangular measurements (Hasekamp and Landgraf, 2007; Mishchenko et al., 2007; Lebsock et al., 2007). Such measurements are rarely performed at the moment although they are highly desirable, if one needs to meet requirements on the accuracy of retrievals needed for the monitoring of aerosol forcing from space (Mishchenko et al., 2004).

Aerosol properties over land and ocean have mainly been retrieved using passive spectral reflectance measurements in the past. Some instruments like Advanced Along-Track Scanning Radiometer (AATSR) onboard Environmental Satellite (ENVISAT) have the capability of a dual-view of the same scene (http://envisat.esa.int/instruments/aatsr/). This constrains the solution of the inverse solution in a much better way than just spectral reflectance measurements. In particular, the forward view of AATSR (55 degrees from normal) senses the atmospheric aerosols on much higher paths than nadir observations, making retrievals more sensitive to small aerosol loads. The POLDER instrument (Deschamps et al., 1994) has the capability of measuring the Stokes vector of reflected light in addition to multiple views of the same ground scene. This enhances retrievals considerably, especially over land. This is because the surface reflection in the polarized light changes little. Therefore, it can be estimated using IR measurements, where the aerosol contribution is usually low. It was found by Lebsock et al. (2007) that polarization observations in retrievals over land not only constrain microphysical properties but also reduce the error in the retrieved aerosol optical thickness.

Several algorithms have been applied to satellite datasets to solve the inverse problem of separating the surface and atmospheric scattering contributions. For instance, MODIS retrievals of aerosol over land (Kaufman et al., 1997; Remer et al., 2005) are based on the correlation of reflectances in the visible and shortwave infrared (SWIR). In essence, the algorithm assumes that the influence of aerosols on the top-of-atmosphere (TOA) reflectance in the SWIR is negligible. Therefore, the ground surface reflectance can be found at these wavelengths, (e.g., at 2.1 μm for MODIS) by only correcting for Rayleigh scattering and gaseous absorption in the atmosphere. One can then exploit the correlation between the SWIR ground reflectances with those in the visible channels where aerosol scattering is significant. The derived surface reflectance is used for constraining the aerosol retrievals. Another possibility is to use multi-angle observations of the same ground scene, as is done with MISR and AATSR (Diner et al., 2005; Grey et al., 2006a,b; North et al., 1999, 2002). This makes it possible to accurately account for directional surface scattering in the retrieval procedure. Some studies use polarized light for aerosol retrieval, e.g., from POLDER, employing the fact that atmospheric scattering is much more polarized than surface reflection (Deuze et al., 2001). The BAER-MERIS algorithm (von Hoyningen-Huene et al., 2003) is based on studies of TOA reflectances in the blue region, where most surface types are only weakly reflective and the scattering from the atmosphere contributes more to the observed signal (Hsu et al., 2004, 2006).

However, these diverse algorithms and approaches do not always give consistent values of the aerosol properties for a given ground scene. The problem is further complicated by the fact that the information content of satellite measurements is underconstrained as far as aerosol measurements are concerned. It is not always possible to constrain the phase function and the single-scattering albedo from measurements themselves. Therefore, *a priori* assumptions are used that are typically based on prescribed aerosol models. Depending on the aerosol properties employed, and on the performance of the algorithms and accuracy of the underlying assumptions, different values of aerosol optical thickness may be retrieved.

5.2.2 Passive satellite instruments: an overview

Currently, satellite-based AOT retrieval techniques are developed by different research teams. A range of algorithms has been designed because the satellite sensors have different characteristics in terms of temporal, spatial, polarization, angular and spectral information content. Although these retrieval algorithms are different, they should ideally produce consistent values for the aerosol properties for a given scene. The characteristics of selected satellite instruments are shown in Table 5.2. MERIS, AATSR, and SCIAMACHY are on ENVISAT, MISR and MODIS are installed on TERRA, and POLDER is onboard PARASOL. AATSR, MERIS, and SCIAMACHY can be compared directly because they measure at the same place and time, so in theory retrievals of AOT should be consistent across these three instruments. AOTs derived from instruments onboard different platforms may differ because there is a time difference between the observations. For instance, TERRA flies by approximately 30 minutes later than ENVISAT, and PARASOL approximately 90

Table 5.2. The characteristics of selected satellite instruments

Instrument	Satellite/ time of measurement	Swath	Channels	Spatial resolution	Multi-angle observation
MERIS	ENVISAT 10:00UTC	1150 km	15 bands 0.4–1.05 µm (0.41, 0.44, 0.49, 0.51, 0.56, 0.62, 0.665, 0.681, 0.705, 0.754, 0.76, 0.775, 0.865, 0.89, 0.9 µm)	0.3×0.3 km^2	no
AATSR	ENVISAT 10:00UTC	512 km	7 bands 0.55, 0.66, 0.87, 1.6, 3.7, 10.85, 12.0 µm	1×1 km^2	yes, two angles from the ranges 0–21.732 and 55.587–53.009 degrees
SCIAMA-CHY	ENVISAT 10:00UTC	916 km	8000 spectral points 0.24–2.4 µm	30×60 km^2	no
MISR	TERRA 10:32UTC	400 km	4 bands 0.446, 0.558, 0.672, 0.866 µm	0.25×0.25 km^2 at nadir and at 0.672 µm 1.1×1.1 km^2 in the remaining channels	yes, nine angles 0, 26.1, 45.6, 60.0, 70.5 degrees
MODIS	TERRA 10:32UTC AQUA 13:30UTC	2300 km	36 bands 0.4–14.4 µm (1): 0.659, 0.865 (2): 0.47, 0.555, 1.24, 1.64, 2.13 (3): 0.412, 0.443, 0.488, 0.531, 0.551, 0.667, 0.678, 0.748, 0.869, 0.905, 0.936, 0.94, 1.375 + MWIR (6)/LWIR (10) channels	(1): 0.25×0.25 km^2 (2): 0.5×0.5 km^2 (3): 1×1 km^2	no
POLDER	PARASOL 13:33UTC	1700 km	8 bands 0.443, 0.490*, 0.565, 0.670*, 0.865*, 0.763, 0.765, 0.91	5.3×6.2 km^2	yes (for channels denoted by '*')

minutes after ENVISAT. Therefore, AOTs derived from MISR, MODIS, and especially POLDER may not be identical to those obtained from instruments on ENVISAT. In addition, the standard POLDER algorithm derives only the fine fraction contribution and not the total AOT.

One problem arises due to different the spatial resolutions of different instruments (see Table 5.2). MODIS performs measurements with the spatial resolution of 0.5×0.5 km^2 at 0.55 µm, which is somewhat larger than those of MERIS (0.3×0.3 km^2). MISR radiance data are acquired at 0.275×0.275 km^2 and 1.1×1.1 km^2, depending on channel, and aerosol products are derived at 17.6×17.6 km^2 resolution. AATSR has a resolution of 1×1km^2 and POLDER has the spatial resolution 5.3×6.2 km^2. The time of acquisition of the instruments is given in Table 5.2.

5.2.3 Determination of aerosol optical thickness from space

Because the characteristics of the satellite instruments differ, algorithms for aerosol retrieval have tended to be sensor-specific. For some instruments several algorithms have been developed. In this section an overview of the different algorithms is given. The characteristics of the datasets are summarized in Table 5.3.

MERIS

Two MERIS algorithms are available and incorporated in the standard software distributed by the European Space Agency. The first algorithm was developed by Santer et al. (1999, 2000) specifically for aerosol retrievals from the MERIS instrument. The results of these retrievals are routinely distributed by ESA as a standard product. The ESA MERIS algorithm is based on the look-up table (LUT) approach for selected aerosol size distributions with given refractive indices. It is assumed that particles have a spherical shape and the reflection from the ground is low. The algorithm fails in the cases of bright ground or nonspherical scatterers (e.g., desert dust aerosols). A detailed description is given in the MERIS Algorithm Theoretical Basis Document (ATBD) 2.15 (Santer et al., 2000). In practice, the ESA MERIS algorithm consists of two different routines, depending on the underlying surface. In both cases the retrieval relies on the knowledge of the underlying surface. Over water, two bands in the near-infrared (NIR) (0.779 µm and 0.865 µm) and in the green (0.51 µm) are used. Over land, two bands in the blue (0.412 µm and 0.443 µm) and one in the red (0.665 µm) are used. Starting from the top-of-atmosphere reflectance, first a gaseous correction is performed with ozone as auxiliary data. The sur-

Table 5.3. Instruments and aerosol retrieval algorithms

No.	Instrument	Algorithm	Reference	Spatial resolution of reported AOT	Remarks
1.	MERIS	ESA	Santer et al. (1999)	1×1 km^2	Standard ESA product
2.	MERIS	BAER	von Hoyningen-Huene et al. (2003)	1×1 km^2	NDVI-based retrievals
3.	AATSR	AATSR-1	Grey et al. (2006b)	10×10 km^2	Dual-view technique
4.	AATSR	AATSR-2	Thomas et al. (2007)	3×3 km^2	Dual-view technique
5.	AATSR	AATSR-3	Thomas et al. (2007)	3×3 km^2	Single-view technique
6.	SCIAMACHY	ASP	Di Nicolantonio et al. (2006)	30×30 km^2	Single view hyperspectral measurements
7.	MISR	JPL	Diner et al. (2005)	17.6×17.6 km^2	Multiple view technique
8.	MODIS	NASA	Kaufman et al. (1997)	10×10 km^2	Spectral correlation technique
9.	MODIS	MBAER	Lee et al. (2005)	1×1 km^2	AFRI-based retrievals
10.	POLDER	CNES	Deuze et al. (2001)	5.3×6.2 km^2	Multiple-angle polarized light measurements (16 angles, up to 50° cross-track and up to 60° along-track)

face pressure is determined from the oxygen absorption. Auxiliary data are the surface pressure at sea level and a digital elevation model. The apparent reflectance is then corrected for Rayleigh scattering. In the algorithm, aerosol parameters are retrieved based on comparisons of measured radiances with pre-calculated look-up tables for a representative set of aerosol models. Details on the aerosol models are given by Santer et al. (1999, 2000). The atmospherically resistant vegetation index (Kaufman and Tanre, 1992) is then used to detect the dark dense vegetation pixels for land aerosol remote sensing. An auxiliary dataset, which is provided by POLDER, gives bi-directional reflectance versus time and location. The last module retrieves the aerosol optical thickness at 0.443 µm and the Angstrom exponent. The MERIS standard aerosol product is also processed by the French company ACRI-ST, and supported by the ESA GSE project PROMOTE.

The second algorithm used for MERIS is BAER (Bremen AErosol Retrieval) which was developed by von Hoyningen-Huene et al. (2003). The algorithm is used by ESA for atmospheric correction of the MERIS land surface product. The algorithm is incorporated in ESA BEAM Toolbox (http://www.brockmann-consult.de/beam/). BEAM is the Basic ERS & Envisat (A)ATSR and Meris Toolbox and is a collection of executable tools and an application programming interface, which have been developed to facilitate the utilization, viewing and processing of ESA MERIS, (A)ATSR and ASAR data. The purpose of BEAM is not to duplicate existing commercial packages, but to complement them with functions dedicated to the handling of Envisat MERIS and AATSR products.

Although BAER is similar to the algorithm developed by Santer et al. (1999), it has special LUTs based on the experimentally measured phase function valid for central Europe. The main steps for the determination of the aerosol reflectance in the framework of BAER are:

- the determination of the spectral TOA reflectance for the selected bands using satellite data;
- the subtraction of the Rayleigh path reflectance for the geometric conditions of illumination and observation within the pixel;
- the estimation of the spectral surface reflectance for land and ocean surfaces by linear mixing of different basic spectra (von Hoyningen-Huene et al., 2003) with the coefficient of mixing determined in the NDVI-type approach using wavelengths 0.665 and 0.865 µm;
- smoothing the retrieved spectral AOT, using an Angstrom power law, by means of the iterative modification of the apparent surface reflectance. The retrieved aerosol reflectance is then used to derive the aerosol optical thickness applying corresponding LUTs obtained by radiative transfer modeling.

AATSR

North et al. (1999) developed a simple physical model of light scattering that is pertinent to the dual-angle sampling of the AATSR instrument and can be used to separate the surface bi-directional reflectance from the atmospheric aerosol properties without recourse to *a priori* information on the land surface properties (AATSR-1 algorithm in Table 5.3). Studies have shown that the angular variation of bi-directional reflectance at the different optical bands of the ATSR-2 and AATSR instruments are similar (e.g. Veefkind et al., 1998, 2000). North et al. (1999) add to this work by considering the variation of the diffuse fraction of light with wavelength, where scattering by atmospheric aerosols tends to be greater at shorter wavelengths. This is important because the fraction of diffuse to direct radiation influences the anisotropy of light reflectance from the surface. Considering these contributions results in a physical model of spectral change with view angle (North et al., 1999). To constrain the inverse problem so that AOT is the only unknown atmospheric parameter, assumptions are made concerning the other aerosol optical properties (e.g. phase function and single-scattering albedo). The algorithm uses pre-calculated look-up tables derived from the 6S (Second Simulation of the Satellite Signal in the Solar Spectrum) radiative transfer model of Vermote et al. (1997) to allow for rapid inversion. A numerical iteration is used to search through different atmospheric profiles to find the AOT that results in the optimal set of surface reflectances.

The retrieved properties include a set of eight surface bi-directional reflectance factors at four wavelengths and two angles, AOT at 0.55 μm and an estimate of the tropospheric aerosol model that falls into one of five compositional categories including continental (predominantly composed of dust-like particles), urban, sea-salts, biomass (smoke) and desert-dust aerosols.

Yet another retrieval algorithm for AATSR was developed at Oxford University (UK). The Oxford-RAL retrieval of Aerosol and Cloud properties, known as ORAC (Thomas et al., 2005) is an optimal estimation (OE) scheme designed for retrievals from near-nadir satellite radiometers. ORAC can be used for dual-view retrievals and also for single-view (in nadir or forward observation mode) retrievals. The AATSR onboard ENVISAT and SEVIRI on board METEOSAT aerosol data obtained with this ORAC algorithm, together with the ESA MERIS product, are supported by the ESA DUE Project GLOBAEROSOL (Carboni et al., 2006), and both individual and merged AOT data are available. The acquisition time of SEVIRI is quite high. Therefore, one is able to follow pollution plumes in a way similar to that known for many years from TV broadcasts of cloud fields. This is not possible with the use of polar orbiting satellites such as ENVISAT (ESA) or TERRA (NASA).

ORAC currently retrieves aerosol optical depth at 0.55 μm, effective radius of aerosol particles and surface albedo. The algorithm uses the Levenberg–Marquardt method to fit the simulated radiance to the measurements, minimizing a cost function based on OE techniques (Rodgers, 2000).

The forward model takes into account scattering and absorption by aerosol, gases and Rayleigh scattering. The radiative transfer equation is solved, at each wavelength, with DISORT (Stamnes et al., 1998), using 60 streams with the delta-M approximation (Lenoble, 1985) stored in LUTs.

The atmosphere is modelled with 32 layers as described by the United States standard atmosphere model (McCartney, 1977). Each layer is a mixture of molecules and aerosol and is characterized by a value of optical depth and single-scattering albedo and phase function (expressed in terms of Legendre moments). Aerosol is placed only at height levels appropriate for the aerosol type. Aerosol optical properties are obtained using Mie theory (Grainger et al., 2004). Every aerosol type considered is a combination of different aerosol components (from the OPAC database; Hess et al., 1998) and the mixing ratio is changed in order to obtain different effective radii.

Gas absorption optical depth for the local gases are obtained from MODTRAN (Moderate Resolution Transmittance code) v3.5 and convolved with the instrumental channel spectral shape.

An optimal estimation approach to the retrieval of parameters enables the extraction of information from all channels simultaneously. This method also allows characterization of the error in each parameter in each individual observation (or 'pixel') under the assumption that the aerosol observed is consistent with the modeled aerosol (i.e. reasonably plane-parallel in nature). A second diagnostic (the solution cost) indicates whether in fact this assumption is true. The OE framework also enables the use of any prior information on the pixel observed. In particular, *a priori* information on the surface albedo is used. The scheme uses surface reflectances based on the MODIS BRDF product (Jin et al., 2003) over land and a model based on Cox and Munk (1954) wave slope statistics over ocean. The surface albedo is retrieved by first assuming an albedo spectral shape for the 0.55, 0.67, 0.87 and 1.6 μm channels. The retrieval searches for the solution with the lowest cost by varying the albedo in the 0.55 μm channel and keeping the respective ratios of all other channels to this channel constant.

The dual-view aerosol retrieval is an extension of the scheme described above. Instead of using data from one viewing geometry it uses both forward and nadir measurements simultaneously, and retrieves a pair of surface albedo values instead of one. The treatment of aerosol is the same as in the single-view algorithm. The AATSR dual-view retrieval algorithm is only carried out on pixels where both forward and nadir data is not flagged as cloud-contaminated (using a ratio threshold test).

A variable resolution of AOT product is possible by averaging nearby data points. Often the retrieval is performed at 3×3 km^2 resolution, this 'superpixelling' of data decreases the effective noise of the measurements by a factor of up to 3 for a completely cloud-free superpixel.

MISR

The MISR retrieval technique uses measurements at nine angles and four spectral bands to constrain the aerosol retrievals. The MISR algorithm makes use of a prescribed set of aerosol models considered to be representative of the types to be found over the globe, and determines for which models, and at what optical depth for each model, a set of acceptance criteria is satisfied. The models are bimodal or trimodal mixtures of fine mode aerosols of various size distributions and single-scattering albedo, coarse-mode aerosols, and nonspherical dust. Air mass factors ranging from 1 to 3 (owing to the view angle range from nadir to 70°) provide considerable sensitivity to aerosol optical depth, especially for thin haze. MISR's nine near-simultaneous views also cover a broad range of scattering angles, between about 60° and 160° in mid-latitudes. Over land, the principal problem is separating the surface and atmospheric contributions to the observed top-of-atmosphere radiances. MISR takes advantage of the increasing ratio of atmospheric to surface contributions to the top-of-atmosphere signal with increasing view zenith angle. The MISR algorithm models the shape of the surface bi-directional reflectance as a linear sum of angular empirical orthogonal functions derived directly from the image data, making use of spatial contrast and angular variation in the observed signal to separate the surface and atmospheric signals, even in situations where bright, dusty aerosols overlie a bright, dusty surface (Martonchik et al., 2002; Diner et al., 2005). A constraint imposing spectral invariance in the angular shape of the surface directional reflectance is also employed in the retrievals (Diner et al., 2005). Globally, MISR optical depths have been validated against AERONET and other Sun photometers over a wide variety of surface types (e.g., Martonchick et al., 2004; Abdou et al., 2004; Kahn et al., 2005).

MODIS

Two MODIS AOT retrieval techniques have been published. One is based on the NASA near IR-visible surface albedo correlation approach (Kaufman et al., 1997; Levy et al., 2007a,b) and the other is the modified BAER (MBAER) algorithm described by Lee et al. (2005, 2006).

Operational aerosol product of MODIS level 2 aerosol datasets (MOD04 L2; MODIS aerosol product, Version 4.1.3) obtained using the technique described by Kaufman et al. (1997b) can be collected from National Aeronautics and Space Administration Distributed Active Archive Center (http://eosdata.gsfc.nasa.gov/). The retrieval is based on the fact that the aerosol contribution is low at 2.1 µm. This enables an accurate determination of the surface contribution at this wavelength. The information on the surface reflectance in the near-infrared is used to estimate the surface reflectance in the visible (Kaufman et al., 1997b). The MOD04 data has various aerosol physical and optical parameters with 10×10 km^2 spatial resolution. The MODIS AOT has been validated with ground-based Sun photometer AOT by a spatio-temporal approach (Ichoku et al., 2002). It has been shown that the MODIS aerosol retrievals over land surface, except in coastal zones, are found within retrieval errors $\Delta AOT = \pm 0.05 \pm 0.2 AOT$ (Chu et al., 2002).

MBAER (Lee et al., 2005, 2006) uses a so-called Aerosol Free Vegetation Index (AFRI; Karnieli et al., 2001) coupled with LUTs constructed using SBDART code (Ricchiazzi et al., 1998) for 1-km resolution MODIS AOT retrieval. The clouds and Sun glint pixels are masked using the MODIS clear sky discriminating method (MOD35; Ackerman et al., 1998). The Rayleigh optical thickness is obtained (Buchholtz, 1995) from the surface pressure determined by the height of the ground for the pixel analyzed. The separation of surface reflectance from TOA reflectance over land involves the use of a linear mixing model of the spectral reflection of green vegetation and soil. The spectra used are given by von Hoyningen-Huene et al. (2003). This enables the estimation of the land surface reflectance for a given pixel. For the contribution of vegetation spectra, the aerosol-free vegetation index (Karnieli et al., 2001) is used. Since this index can minimize aerosol effect, the vegetation fraction can be determined quite accurately. Surface reflectance determined by the linear mixing model tuned by the corrected aerosol-free vegetation index is then used to determine AOT.

POLDER

POLDER (POLarization and Directionality of Earth Reflectances; Deschamps et al., 1994) performs multi-angle measurements of the sunlight reflected by the Earth surface and atmosphere at eight spectral bands in the visible and near-infrared spectral domain (0.443 to 1.02 µm). There is the third version of the instrument onboard the micro-satellite PARASOL, while the two previous versions were onboard ADEOS 1 and 2. Multi-directional and polarization measurements provide additional information to retrieve aerosol load in the atmosphere. Indeed, the reflectance from the surface shows little polarization while that of fine aerosol is highly polarized. As a consequence, the relative contribution of the aerosols to the top-of-the-atmosphere reflectance is much higher for the polarized component than that for the total component, which makes it easier to identify aerosol signatures than with the other instruments discussed here. On the other hand, coarse aerosols generate little polarized light so that the POLDER retrieval focuses on the fine fraction of the aerosols. Aerosol generated by pollution and biomass burning are mostly in the fine mode and are therefore well captured by the retrieval method (Deuze et al., 2001). On the other hand, dust is mostly in the coarse mode. The retrieval algorithm assumes spherical scatterers, which is valid for fine aerosols. The contribution from the surface to the polarized reflectance is based on *a priori* values (as a function of observation geometry and surface type) derived from a statistical analysis of POLDER data (Nadal et al., 1999). The aerosol load and type is obtained through a classical LUT algorithm based on the multi-directional polarized reflectance measurements at 0.67 and 0.865 µm.

5.2.4 Spatial distribution of aerosol optical thickness

The browse image of the cloudless scene over Germany on October 13, 2005, is shown in Fig. 5.13. Results of AOT retrievals for this scene using MODIS Collection 5 data are given in Fig. 5.14(a) and retrievals using MISR are shown in Fig. 5.14(b). Both retrievals are consistent and indicate pollution (green colour) seen on the general pattern of background aerosol (blue color). The latitude range of retrievals is from 49 N to 53 N and the longitude range is from 7 E to 12 E. The surface is mostly covered by dense vegetation with some areas of bare soil. The surface is not as bright as compared, for example, to dry regions such as those in southern Europe and the Sahara. This makes aerosol retrievals less dependent on surface reflective characteristics (e.g., bi-directional reflection distribution function).

There is some indication of small clouds in some parts of the study area which is, however, mostly cloud-free (see Fig. 5.13). The humidity was low (below 40 % for most of area) and the boundary layer height was about 1 km, as indicated by ECMWF analysis. The analysis of relevant meteorological data suggests that the situation was characterized

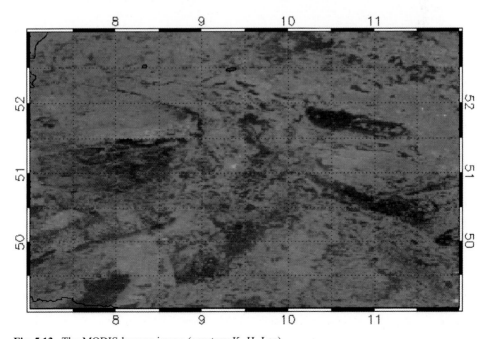

Fig. 5.13. The MODIS browse image (courtesy K.-H. Lee).

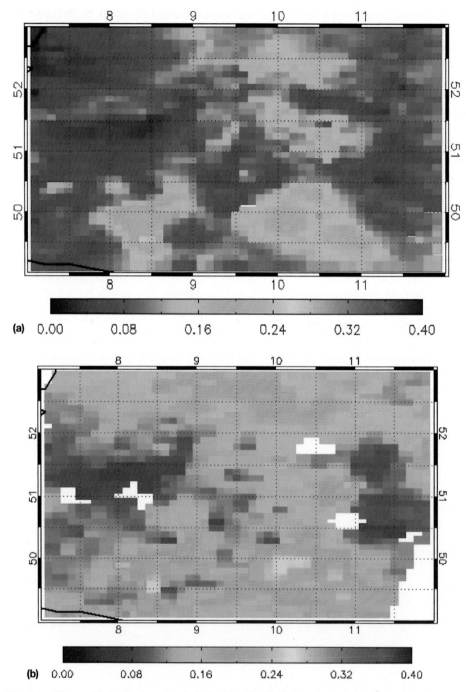

Fig. 5.14. The spatial distribution of AOT derived from MODIS (a) and MISR (b) observations (courtesy K.-H. Lee).

by high-pressure conditions with the average temperature of 14 °C at 10:00UTC and 17 °C at 13:00UTC.

Several AERONET (Holben et al., 1998) instruments operated at the time of the satellite measurements. Their locations, the time of measurements, and the values of aerosol optical thickness at wavelengths 0.44, 0.55, and 0.67 μm are given in Table 5.4. It follows from this table that the average AOT at 0.55 μm for most Sun photometers was close to 0.2 on October 13, 2005, for central Europe. It is close to the value derived from satellite measurements given in Figs 5.13 and 5.14 and also in Table 5.5. In Fig. 5.15, we compare AERONET AOT measurements with those retrieved using different algorithms described above. It follows that most algorithms perform quite well for the area studied, where the cloud contamination was very small or absent. In practice, however, the identification of cloudy pixels is not trivial, especially for very thin cirrus clouds. Currently, comprehensive cloud-screening algorithms are under development. They include temporal and spatial variability tests, the screening of most dark (cloud shadows) and most bright pixels in a given area, spectral ratios (e.g., UV/visible), thermal infrared measurements, and also measurements at spectral intervals, where atmospheric gases (e.g., water vapor, oxygen) have absorption bands.

Table 5.4. The AERONET aerosol optical thickness at 0.44 μm and 0.67 μm obtained at several locations in Europe on September 13, 2005 (Bremen is not an official AERONET site). The value of aerosol optical thickness τ at 0.55 μm was obtained using corresponding Angstrom coefficients (Kokhanovsky et al., 2007)

Station	Position	Time	$\tau(0.44 \ \mu m)$	$\tau(0.55 \ \mu m)$	$\tau(0.67 \ \mu m)$
Hamburg	53.568 N, 9.973 E	09:52	0.21	0.15	0.11
Helgoland	54.178 N, 7.887 E	09:45	0.27	0.20	0.15
Cabauw	51.971 N, 4.927 E	09:57	0.25	0.19	0.15
Den Haag	52.110 N, 4.327 E	09:44	0.31	0.22	0.16
Leipzig	51.354 N, 12.435 E	10:06	0.24	0.17	0.13
Mainz	49.999 N, 8.300 E	09:58	0.42	0.31	0.24
Karlsruhe	49.093 N, 8.428 E	09:43	0.31	0.22	0.16
ISGDM_CNR	45.437 N, 12.332 E	09:59	0.57	0.41	0.31
Venice	45.314 N, 12.508 E	09:29	0.47	0.41	0.24
Bremen	53.05 N, 8.78 E	10:06	0.35	0.26	0.20

Table 5.5. Statistical characteristics of retrieved AOT at 0.55 μm for the area 9–11.5 E, 52–52.5 N (October 13, 2005) (Kokhanovsky et al., 2007)

Instrument/algorithm	Average AOT	Standard deviation
MODIS/NASA	0.15	0.03
MISR/JPL	0.16	0.02
POLDER	0.16	0.04
MERIS/BAER	0.20	0.02
MODIS/BAER	0.20	0.01
MERIS/ESA	0.21	0.05
AATSR/ORAC	0.22	0.06

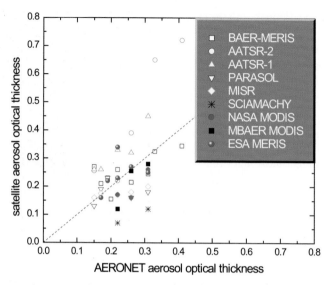

Fig. 5.15. The comparison of satellite and ground measurements of the aerosol optical thickness at the wavelength 0.55 μm (Kokhanovsky et al., 2007). AATSR-1 retrievals correspond to the ORAC algorithm. AATSR-2 retrievals correspond to the North et al. (1999) algorithm. SCIAMACHY AOT is obtained using the di Nicolantonio et al. (2006) algorithm (Kokhanovsky et al., 2007).

5.2.5 Lidar sounding from space

Passive measurements do not provide accurate information on the detailed aerosol vertical distribution. But some results can be obtained using aerosol profiling in the gaseous absorption bands and also with stereoscopy for thick aerosol layers. Lidar measurements from space are ideal for such investigations especially for cloud-free conditions, when there is no disturbance due to overlying clouds.

The type of information, which can be obtained is demonstrated in Fig. 5.16, where data of the Cloud-Aerosol Lidar with Orthogonal Polarization (CALIOP) onboard CALIPSO spacecraft are presented. CALIOP measures at 532 nm (parallel and perpendicular polarizations) and also at 1064 nm. The pulse repetition rate is 20.16 Hz, pulse length is 20 ns, and the pulse energy is 110 mJ. The receiver FOV is 130 μrad. The vertical resolution of CALIOP is in the range 30–300 m with a horizontal resolution of 330–5000 m depending on the altitude range.

Fig. 5.16 confirms that CALIOP allows for an easy identification of dust plumes and biomass smoke. Particles in smoke are roughly spheres and also they are quite small, therefore, they give almost no signal in the 532-nm perpendicular attenuated backscatter. This is not the case for desert dust nor for cirrus clouds, where particles are large and irregularly shaped.

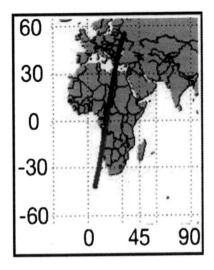

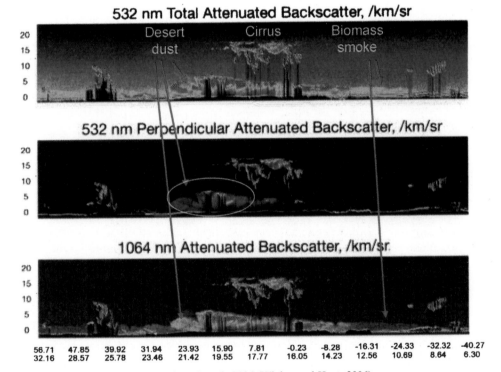

Fig. 5.16. CALIOP observations from June 9, 2006 (Winker and Hunt, 2006).

Appendix: Spectral dependence of Rayleigh optical thickness and depolarization factor

Table A1. The dependence of Rayleigh optical thickness on the wavelength λ (the first column, in µm for different atmospheric models: A, tropical (15 N); B, midlatitude summer (45 N, July); C, midlatitude winter (45 N, January); D, subarctic summer (60 N, July); E, subarctic winter (60 N, January); F, 1962 US standard model (Bucholtz, 1995)

λ	A	B	C	D	E	F
0.20	7.819	7.807	7.826	7.761	7.783	7.788
0.21	6.139	6.129	6.145	6.093	6.111	6.114
0.22	4.912	4.904	4.917	4.876	4.890	4.892
0.23	3.986	3.979	3.990	3.956	3.968	3.970
0.24	3.279	3.274	3.282	3.255	3.264	3.266
0.25	2.725	2.721	2.728	2.705	2.713	2.714
0.26	2.286	2.282	2.288	2.269	2.276	2.277
0.27	1.935	1.932	1.937	1.920	1.926	1.927
0.28	1.648	1.645	1.650	1.636	1.640	1.641
0.29	1.414	1.412	1.416	1.404	1.408	1.409
0.30	1.221	1.219	1.222	1.212	1.216	1.216
0.31	1.061	1.060	1.062	1.053	1.056	1.057
0.32	9.262	9.247	9.271	9.193	9.220	$9.225*(\times 10^{-1})$
0.33	8.121	8.108	8.129	8.061	8.084	8.088
0.34	7.158	7.147	7.165	7.105	7.126	7.130
0.35	6.330	6.320	6.336	6.283	6.301	6.304
0.36	5.624	5.615	5.629	5.582	5.598	5.601
0.37	5.014	5.006	5.019	4.977	4.991	4.994
0.38	4.482	4.475	4.486	4.449	4.461	4.464
0.39	4.022	4.016	4.026	3.992	4.004	4.006
0.40	3.620	3.615	3.624	3.594	3.604	3.606
0.41	3.267	3.262	3.270	3.243	3.252	3.254
0.42	2.956	2.952	2.959	2.935	2.943	2.945
0.43	2.682	2.678	2.684	2.662	2.670	2.671
0.44	2.439	2.435	2.441	2.421	2.428	2.429
0.45	2.223	2.219	2.225	2.206	2.212	2.214
0.46	2.030	2.027	2.032	2.015	2.021	2.022
0.47	1.858	1.855	1.860	1.844	1.849	1.850
0.48	1.704	1.701	1.705	1.691	1.696	1.697
0.49	1.565	1.563	1.567	1.554	1.558	1.559
0.50	1.441	1.438	1.442	1.430	1.434	1.435
0.51	1.329	1.326	1.330	1.319	1.322	1.323
0.52	1.227	1.225	1.228	1.218	1.222	1.222
0.53	1.135	1.133	1.136	1.127	1.130	1.131
0.54	1.052	1.050	1.053	1.044	1.047	1.048
0.55	9.760	9.745	9.769	9.688	9.761	$9.721**(\times 10^{-2})$
0.56	9.067	9.053	9.076	9.000	9.026	9.031

Table A1. *(cont.)*

λ	A	B	C	D	E	F
0.57	8.435	8.422	8.443	8.373	8.396	8.401
0.58	7.857	7.845	7.865	7.799	7.821	7.826
0.59	7.328	7.316	7.335	7.274	7.294	7.298
0.60	6.842	6.832	6.849	6.792	6.811	6.815
0.61	6.398	6.388	6.404	6.351	6.369	6.372
0.62	5.989	5.980	5.995	5.945	5.962	5.965
0.63	5.612	5.604	5.618	5.571	5.587	5.590
0.64	5.265	5.257	5.270	5.226	5.241	5.244
0.65	4.944	4.936	4.949	4.908	4.922	4.924
0.66	4.647	4.640	4.652	4.613	4.626	4.629
0.67	4.373	4.366	4.377	4.340	4.353	4.355
0.68	4.118	4.112	4.122	4.088	4.099	4.102
0.69	3.882	3.876	3.885	3.853	3.864	3.866
0.70	3.662	3.656	3.666	3.635	3.645	3.647
0.71	3.458	3.452	3.461	3.432	3.442	3.444
0.72	3.268	3.262	3.271	3.243	3.253	3.254
0.73	3.090	3.085	3.093	3.067	3.076	3.078
0.74	2.925	2.920	2.928	2.903	2.912	2.913
0.75	2.770	2.766	2.773	2.750	2.758	2.759
0.76	2.626	2.621	2.628	2.606	2.614	2.615
0.77	2.490	2.486	2.492	2.472	2.479	2.480
0.78	2.363	2.359	2.365	2.346	2.352	2.354
0.79	2.244	2.241	2.246	2.228	2.234	2.235
0.80	2.133	2.129	2.135	2.117	2.123	2.124
0.90	1.326	1.324	1.328	1.317	1.320	1.321
1.0	8.680	8.666	8.688	8.616	8.640	$8.645{***}(\times 10^{-3})$
1.10	5.917	5.908	5.923	5.873	5.890	5.893
1.20	4.171	4.165	4.175	4.141	4.152	4.155
1.30	3.025	3.020	3.028	3.003	3.011	3.013
1.40	2.247	2.243	2.249	2.230	2.237	2.238
1.50	1.704	1.701	1.705	1.691	1.696	1.697
1.60	1.315	1.313	1.317	1.306	1.309	1.310
1.70	1.032	1.030	1.033	1.024	1.027	1.027
1.80	8.204	8.191	8.212	8.143	8.167	$8.171{****}(\times 10^{-4})$
1.90	6.606	6.596	6.613	6.557	6.576	6.580
2.00	5.379	5.371	5.384	5.339	5.355	5.358
2.20	3.672	3.666	3.676	3.645	3.655	3.658
2.40	2.592	2.588	2.594	2.573	2.580	2.581
2.60	1.881	1.878	1.883	1.867	1.873	1.874
2.80	1.398	1.396	1.400	1.388	1.392	1.393
3.00	1.061	1.059	1.062	1.053	1.056	1.057
3.50	5.724	5.715	5.730	5.682	5.698	$5.702{*****}(\times 10^{-5})$
4.00	3.355	3.350	3.358	3.330	3.340	3.341

Table A2. The spectral dependence of the depolarization factor ρ (Bucholtz, 1995). The phase function of Rayleigh scattering is given by the following expression:

$$p(\theta) = \frac{3(1 + 3\gamma + (1 - \gamma)\cos^2\theta)}{4(2 + 2\gamma)}, \quad \gamma = \frac{\rho}{2 - \rho}$$

λ, μm	$100\,\rho$
0.2	4.545
0.205	4.384
0.21	4.221
0.215	4.113
0.22	4.004
0.225	3.895
0.23	3.785
0.24	3.675
0.25	3.565
0.26	3.455
0.27	3.4
0.28	3.289
0.29	3.233
0.3	3.178
0.31	3.178
0.32	3.122
0.33	3.066
0.34	3.066
0.35	3.01
0.36	3.01
0.37	3.01
0.38	2.955
0.39	2.955
0.4	2.955
0.45	2.899
0.5	2.842
0.55	2.842
0.6	2.786
0.65	2.786
0.7	2.786
0.75	2.786
0.8	2.73
0.85	2.73
0.9	2.73
0.95	2.73
1	2.73

References

Abdou, W.A., D.J. Diner, J.V. Martonchik, et al., 2004: Comparison of coincident MISR and MODIS aerosol optical depths over land and ocean scenes containing AERONET sites, J. Geophys. Res., 110, D10S07, doi:10.1029/2004JD004693.

Ackerman, S.A., et al., 1998: Discriminating clear-sky from cloud with MODIS. Algorithm theoretical basis document (MOD35), J. Geophys. Res., 103, 32141–32157.

Ambartsumian V.A., 1943: On scattering of light by a diffuse medium, Dokl. Akad. Nauk SSSR 38, 257–265 [Compt. Rend. (Doklady) Acad. Sci. USSR 38, 229–232 (1943)].

Ambartsumian, V.A., 1961: *Scientific Papers*, vol. 1, Erevan: Armenian Academy of Sciences.

Anderson, E., et al., 1995: *LAPACK. User's Guide*, 2nd Edition. Philadelphia: SIAM.

Andrews, E., et al., 2006: Comparison of methods for deriving aerosol asymmetry parameter, J. Geophys. Res., 111, D05S04,10.1029/2004JD005734.21.

Angstrom, A., 1929: On the atmospheric transmission of sun radiation and on the dust in the air, Geogr. Annal., 2, 156–166.

Anikonov, A.S., and S.Yu. Ermolaev, 1977: On diffuse light reflection from a semi-infinite atmosphere with a highly extended phase function, Vestnik LGU, 7, 132–137.

Babenko, V.A., et al., 2003: *Electromagnetic Scattering in Disperse Media: Absorption by Inhomogeneous and Anisotropic Aerosol Particles*, Berlin: Springer–Praxis.

Berry, M.V., and I.C. Percival, 1986: Optics of fractal clusters such as smoke, Opt. Acta, 33, 577–591.

Bohren, C.F., and D.R. Huffman, 1983: *Absorption and Scattering of Light by Small Particles*, New York: Wiley.

Bovensmann, H., et al., 1999: SCIAMACHY: mission objectives and measurement modes, J. Atmos. Sci., 56, 127–150.

Breon, F.-M., D. Tanre, and S. Generoso, 2002: Aerosol effect on cloud droplet size monitored from satellite, Science, 295, 834–838.

Buchholtz, A., 1995: Rayleigh-scattering calculations for the terrestrial atmosphere, Appl. Opt., 34, 2765–2773.

Cadle, R.D., 1966: *Particles in the Atmosphere and Space*, New York: Reinhold.

Carboni E., G. Thomas, D. Grainger, et al., 2006: GlobAEROSOL from Earth Observation – aerosol maps from (A)ATSR and SEVIRI, *CD-ROM Proc. Atmos. Sci. Conference*, Frascati, May 8–12.

Chahine, M.T., 1968: Determination of temperature profile in an atmosphere from its outgoing radiance, J. Opt. Soc. Am., 12, 1634–1637.

Chamaillard K., S.G. Jennings, C. Kleefeld, D., et al., 2003: Light backscattering and scattering by non-spherical sea-salt aerosols, J. Quant. Spectr. Rad. Transfer, 79–80, 577–597.

Chandrasekhar S., 1950: *Radiative Transfer*, Oxford: Oxford University Press.

Chowdhary, J., B. Cairns, M.I. Mishchenko, et al., 2005: Retrieval of aerosol scattering and absorption properties from photopolarimetric observations over the ocean during the CLAMS experiment, J. Atmos. Sci., 62, 1093–1117.

Chu, D.A., Kaufman, Y.J., Ichoku, C., et al., 2002: Validation of MODIS aerosol optical depth retrieval over land, Geophys. Res. Lett., 29, 10.1029/2001GL013205.

Clarke, A., V. Kapustin, S. Howell, et al., 2003: Sea-salt size distributions from breaking waves: implications for marine aerosol production and optical extinction measurements during SEAS, 20, 1362–1374.

Coulson, K.L., J.V. Dave, and Z. Sekera, 1960: *Tables Related to Radiation Emerging from a Planetary Atmosphere with Rayleigh Scattering*, Berkeley: University of California Press.

Cox, C., and W. Munk, 1954: Measurement of the roughness of the sea surface from photographs of the sun's glitter, J. Opt. Soc. Am., 44, 838–850.

Crutzen, P., 2006: Albedo enhancement by stratospheric sulfur injections: a contribution to resolve a policy dilemma? Climatic Change, 77, 3–4, 211–220.

d'Almeida, G.A., et al., 1991: *Atmospheric Aerosols: Global Climotology and Radiative Characteristics*, New York: A. Deepak.

Darwin, Charles, 1845: *Journal of researches into the natural history and geology of the countries visited during the voyage of H.M.S. Beagle round the world, under the Command of Capt. Fitz Roy, R.N.* (2nd edn), London: John Murray.

de Haan, J.F., 1987: *Effects of Aerosols on the Brightness and Polatization of Cloudless Planetary Atmospheres*, PhD thesis, Free University of Amsterdam.

Deirmendjian, D., 1969: *Electromagnetic Scattering on Spherical Polydispersions*, Amsterdam: Elsevier.

de Rooij, W.A., 1985: *Reflection and Transmission of Polarized Light by Planetary Atmospheres*, PhD thesis, Free University of Amsterdam.

Deschamps, P.-Y., F.-M. Bréon, M. Leroy, et al., 1994: The POLDER mission: Instrument characteristics and scientific objectives, IEEE Trans. Geosc. Rem. Sens., 32, 598–615.

Deuze, J.L., F.M. Breon, C. Devaux, et al., 2001: Remote sensing of aerosols over land surfaces from POLDER-ADEOS-1 polarized measurements J. Geophys. Res., 106, 4913–4926.

Diner, D.J., J. Martonchik, R.A. Kahn, et al., 2005: Using angular and spectral shape similarity constrains to improve MISR aerosol and surface retrievals over land, Remote Sens. of Env., 94, 155–171.

Di Nicolantonio, W., A. Cacciari, S. Scarpanti, et al., 2006: Sciamachy TOA reflectance correction effects on aerosol optical depth retrieval, Atmospheric Science Conference, Proceedings of the conference held 8–12 May, 2006 at ESRIN, Frascati, Italy. Edited by H. Lacoste and L. Ouwehand. ESA SP-628. European Space Agency, 2006. Published on CDROM., p. 73.1.

Dubovik, O. and M.D. King, 2000: A flexible inversion algorithm for retrieval of aerosol optical properties from Sun and sky radiance measurements, J. Geophys. Res., 105, 20 673–20 696.

Dubovik, O., B.N. Holben, T.F. Eck, et al., 2002: Variability of absorption and optical properties of key aerosol types observed in worldwide locations, J. Atm. Sci., 59, 590–608.

Dubovik, O., A. Sinyuk, T. Lapyonok, et al., 2006: Application of spheroid models to account for aerosol particle nonsphericity in remote sensing of desert dust, J. Geophys. Res., 111, doi:10.1029/2005JD006619.

Evans, K.F., 1998: The spherical harmonics discrete ordinate method for three-dimensional atmospheric radiative transfer, J. Atmos. Sci., 55, 429–446.

Fiebig, M., J.A. Ogren, 2006: Retrieval and climatology of the aerosol asymmetry parameter in the NOAA aerosol monitoring network, J. Geophys. Res., 111, D21204, doi:10.1029/2005JD006545.

Fock, M.V., 1944: On some integral equations of mathematical physics, Matem. Sbornik, 14, 3–50.

Germogenova, T.A., 1961: On the character of the solution of the transfer equation for a plane layer, Zh. Vych. Mat. I. Mat. Fiz., 1, 1001–1012.

Goloub, P. et al., 2000: Cloud thermodynamical phase classification from the POLDER spaceborne instrument, J. Geophys. Res.: D, 105, 14747–14759.

Gorchakov, G.I., 1966: Light scattering matrices of atmospheric air, Izv. Akad. Nauk SSSR, Fiz. Atmos. Okeana, 2, 595–605.

Gorchakov, G.I., et al., 1976: On the determination of the refraction index of particles using the polarization of light scattered by haze, Izv. Akad. Nauk SSSR, Fiz. Atmos. Okeana, 12, 2, 144–150.

Gordon H.G., 1973: Simple calculation of the diffuse reflectance of ocean. Appl. Opt., 12, 2803–2804.

Grainger, R.G., J. Lucas, G. Thomas, and G. Ewen, 2004: Calculation of Mie derivatives, Appl. Opt., 43, 5386–5393.

Grey, W.M.F., P.R.J. North, and S.O. Los, 2006a: Computationally efficient method for retrieving aerosol optical depth from ATSR-2 and AATSR data, Appl. Opt., 45, 2786–2795.

Grey, W.M.F., P.R.J. North, S.O. Los, R.M. Mitchell, 2006b: Aerosol optical depth and land surface reflectance from multi-angle AATSR measurements: Global validation and inter-sensor comparisons. IEEE Trans. Geosci. Remote Sens., 44, 2184–2197.

Hansen, J.E., and J. Hovenier, 1974: Interpretation of the polarization of Venus, J. Atmos. Sci., 31, 1137–1160.

Hansen, J.E., and L. Nazarenko, 2004: Soot climate forcing via snow and ice albedos, Proc Natl. Acad. Sci., 101, 423–428.

Hansen, J.E., and L.D. Travis, 1974: Light scattering in planetary atmospheres, Space Sci. Rev., 16, 527–610.

Hansen, A.D.A., et al., 1984: The aethalometer – an instrument for the real-time measurement of optical absorption by aerosol particles, The Science of the Total Environment, 36, 191–196.

Hapke, B., 1993: *Theory of Reflectance and Emittance Spectroscopy*, Cambridge: Cambridge University Press.

Hasekamp, O.P., and J. Landgraf, 2007: Retrieval of aerosol properties over land surfaces: capabilities of multiple-viewing-angle intensity and polarization measurements, Appl. Opt., 46, 3332–3344.

Hess, M., P. Koepke, and I. Schult, 1998: Optical properties of aerosols and clouds: the software package OPAC, Bull. Am. Met. Soc., 79, 831–844.

Holben B.N., T.F. Eck, I. Slutsker, et al. 1998: AERONET – A federated instrument network and data archive for aerosol characterization, Rem. Sens. Environ., 66, 1–16.

Horvath, H., 1993: Atmospheric light absorption – A review, Atmos. Environ., 27, 293–317.

Horvath, H., and A. Trier, 1993: A study of the aerosol of Santiago de Chile – I. Light extinction coefficients, Atmospheric Environment. Part A. General Topics, 27, 371–384.

Hovenier, J.W., 1971: Multiple scattering of polarized light in planetary atmospheres, Astron. Astrophys., 13, 7–29.

Hovenier, J.W., C. van Der Mee, and H. Domke, 2004: *Transfer of Polarized Light in Planetary Atmospheres: Basic Concepts and Practical Methods*, (Astrophysics and Space Science Library, vol. 318), Dordrecht: Kluwer Academic.

Hsu, N.C., et al., 2004: Aerosol properties over bright-reflecting source regions, IEEE Trans. Geosci. Rem. Sens., 42, 557–569.

Hsu, N.C., S.C. Tsay, M.D. King, and J.R. Herman, 2006: Deep blue retrievals of Asian aerosol properties during ACE-Asia, IEEE Trans. Geosci. Remote Sens., 44, 3180–3195.

Ichoku C., D.A. Chu, S. Mattoo, et al., 2002: A spatio-temporal approach for global validation and analysis of MODIS aerosol products, Geophys. Res. Lett., 29, doi:10.1029/2001GL013206.

Ignatov, A., and L. Stowe, 2002: Aerosol retrievals from individual AVHRR channels. Part II. Quality control, probability distribution functions, information content, and consistency checks of retrievals, J. Atmos. Sci., 59, 335–362.

IPCC (Intergovernmental Panel on Climate Change), 2001: Climate Change 2001: The Scientific Basis, eds. J.T. Houghton, Y. Ding, D.J. Griggs, et al., Cambridge: Cambridge University Press.

Ishimoto, H., and K. Masuda, 2002: A Monte Carlo approach for the calculation of polarized light: application to an incident narrow beam, J. Quant. Spectr. Rad. Transfer, 72, 467–483.

Jaenicke, R., 1988: Aerosol physics and chemistry, in *Landolt-Bornstein* (edited by S. Bakan et al.), 4b, 391–456, Berlin: Springer.

Jaenicke, R., 2005: Abundance of cellular material and proteins in the atmosphere, *Science*, 308, 73 [DOI: 10.1126/science.1106335] (in Brevia).

Jin, Y., C.B. Schaaf, C.E. Woodcock, et al., 2003: Consistency of MODIS surface BRDF/Albedo retrievals: 1. Algorithm performance: J. Geophys. Res., 108, 4158, doi:10.1029/2002JD002803.

Junge, C.E., 1963: *Air Chemistry and Radiochemistry*, New York: Academic Press.

Kahn, R.A., B.J. Gaitley, K.A. Crean, et al., 2005: MISR global aerosol optical depth validation based on two years of coincident AERONET observations, J. Geophys. Res. 110, D10S04, doi:10.1029/2004JD004706.

Kalashnikova, O.V., and I.N. Sokolik, 2004: Modeling the radiative properties of nonspherical soil-derived mineral aerosols, J. Quant. Spectr. Rad. Transfer, 87, 137–166.

Karnieli, A., Y.J. Kaufman, L.A. Remer, A. Ward, 2001: AFRI – Aerosol free vegetation index, Remote Sensing of Environment, 77, 10–21.

Kaufman, Y., and D. Tanre, 1992: Atmospherically resistant vegetation index (ARVI) for EOS MODIS, IEEE Trans. Geosci. Remote Sens., 30, 261–270.

Kaufman, Y.J., D. Tanre, H.R. Gordon, et al., 1997a: Passive remote sensing of tropospheric aerosol and atmospheric correction for the aerosol effect, J. Geophys. Res., 102, 16,815–16,830.

Kaufman, Y.J., A.E. Wald, L.A. Remer, et al., 1997b: The MODIS 2.1 µm channel-correlation with visible reflectance for use in remote sensing of aerosol, IEEE Trans. Geosci. Remote Sens., 35, 1286–1298.

Kimmel, B.W., and G.V.G. Baranoski, 2007: A novel approach for simulating light interaction with particulate materials: application to the modeling of sand spectral properties, Optics Express, 15, 9755–9777.

King, M.D., Harshvardhan, 1986: Comparative accuracy of selected multiple scattering approximations, J. Atmos. Sci., 43, 784–801.

King, M.D., D.M. Byrne, B.M. Herman, and J.A. Reagan, 1978: Aerosol size distributions obtained by inversion of spectral optical depth measurements, J. Atmos. Sci., 21, 2153–2167.

King, M.D., Y.J. Kaufman, D. Tanre, and T. Nakajima, 1999: Remote sensing of tropospheric aerosols from space: Past, present, and future, Bull. Am. Meteorol. Soc., 80, 2229–2259.

Kinne, S., M. Schulz, C. Textor, et al., 2006: An AeroCom initial assessment optical properties in aerosol component modules of global models, Atmos. Chem. Phys., 6, 1815–1834.

Klett, J.D., 1981: Stable analytical inversion solution for processing lidar returns, Applied Optics 20, 211–220.

Koepke, P., 1984: Effective reflectance of oceanic whitecaps, Appl. Opt., 23, 1816–1824.

Kokhanovsky, A.A., 2003: *Polarization Optics of Random Media*, Berlin: Springer–Praxis.

Kokhanovsky, A.A., 2004a: *Light Scattering Media Optics: Problems and Solutions*, Berlin: Springer–Praxis.

Kokhanovsky, A.A. , 2004b: Reflection of light from nonabsorbing semi-infinite cloudy media: a simple approximation, J. Quant. Spectr. Rad. Transfer, 85, 35–55.

Kokhanovsky, A.A., 2004c: Spectral reflectance of whitecaps, J. Geophys. Res., 109, C05021, doi:10.1029/2003JC002177.

Kokhanovsky, A.A., 2006: *Cloud Optics*, Dordrecht: Springer.

Kokhanovsky A.A., B. Mayer and V.V. Rozanov, 2005: A parameterization of the diffuse transmittance and reflectance for aerosol remote sensing problems, Atmospheric Research, 73, 1–2, 37–43.

Kokhanovsky, A.A., 2007: Scattered light corrections to Sun photometry: analytical results for single and multiple scattering regimes, J. Opt. Soc. America, 24, 1131–1137.

Kokhanovsky, A.A., and T. Nauss, 2006: Reflection and transmission of solar light by clouds: asymptotic theory, Atmos. Chem. Phys. 6, 5537–5545.

Kokhanovsky, A.A., F.-M. Breon, A. Cacciari, et al., 2007: Aerosol remote sensing over land: a comparison of satellite retrievals using different algorithms and instruments, Atmospheric Research, 85, 372–394.

Kuik, F., J.F. de Haan, and J.W. Hovenier, 1992: Benchmark results for single scattering by spheroids, J. Quant. Spectr. Rad. Transfer, 47, 477–489.

Lacis, A.A., and M.I. Mishchenko, 1995: Climate forcing, climate sensitivity, and climate response: A radiative modeling perspective on atmospheric aerosols. In: *Aerosol Forcing of Climate*, Chichester: Wiley.

Lacis, A.A., J. Chowdhary, M.I. Mishchenko, and B. Cairns, 1998: Modeling of errors in diffuse-sky radiation: Vector vs. scalar treatment, Geophys. Res. Lett., 135–138.

Lack, D.A., 2006: Aerosol absorption measurements using photoacoustic spectroscopy: sensitivity, calibration, and uncertainty developments, Aerosol Sci. Technology, 40, 697–708.

Lebsock, M.D., et al., 2007: Information content of near-infrared spaceborne multiangular polarization measurements for aerosol retrievals, J. Geophys. Res., 112, D14206, doi:10.1029/2007JD008535.

Lee, K.H., J.E. Kim, Y.J. Kim, et al. 2005: Impact of the smoke aerosol from Russian forest fires on the atmospheric environment over Korea during May 2003, Atmos. Env., 39, 85–99.

Lee, K.H., Y.J. Kim, W. von Hoyningen-Huene, and J.P. Burrows, 2006: Influence of land surface effects on MODIS aerosol retrieval using the BAER method over Korea, Int. J. Rem. Sens., 27, 2813–2830.

Lenoble, J., ed., 1985: *Radiative Transfer in Scattering and Absorbing Atmospheres: Standard Computational Procedures*, Hampton: A. Deepak.

Levy, R.C., L.A. Remer, S. Mattoo, et al., 2007: Second-generation operational algorithm: Retrieval of aerosol properties over land from inversion of Moderate Resolution Imaging Spectroradiometer spectral reflectance, J. Geophys. Res., 112, D13211, doi:10.1029/2006JD007811.

Lewis E.R. and S.E. Schwartz, 2004: *Sea Salt Aerosol Production: Mechanisms, Methods, Measurements, and Models*, Washington: AGU.

Lin, C.I., M.B. Baker, and R.J. Charlson, 1973: Absorption coefficient of the atmospheric aerosol: a method for measurement, Appl. Opt., 12, 1365–1363.

Liou, K.-N., 2002: *An Introduction to Atmospheric Radiation*, Oxford: Oxford University Press.

Martonchik, J.V., D.J. Diner, K.A. Crean, and M.A. Bull, 2002: Regional aerosol retrieval results from MISR, IEEE Trans. Geosci. Remote Sens., 40, 1520–1531.

Martonchik, J.V., D.J. Diner, R. Kahn, et al., 2004: Comparison of MISR and AERONET aerosol optical depths over desert sites, Geophys. Res. Lett. 31, L16102, doi:10.1029/2004GL019807.

Masuda, K., et al., 1999: Use of polarimetric measurements of the sky over the Ocean for spectral optical thickness retrievals, J. Atmos. Oceanic Techn., 16, 846–859.

McCartney, E.J., 1977: *Optics of the Atmosphere*, New York: Wiley.

McCrone, W.C., 1967: The particle atlas: A photomicrographic reference for the microscopical identification of particulate substances, New York: Ann Arbor Science Publishers.

Meehl, G.A., and C. Tebaldi, 2004: More intense, more frequent, and longer lasting heat waves in the 21st century, Science, 305, 994–997.

Mie, G., 1908: Beiträge zur Optik trüber Medien, speziell kolloidaler Metallösungen. Annalen der Physik, Vierte Folge, 25, 3, 377–445.

Min, Q., and M. Duan, 2004: A successive order of scattering model for solving vector radiative transfer in the atmosphere, J. Quant. Spectr. Rad. Transfer, 87, 243–259.

Mishchenko, M.I., 2002: Vector radiative transfer equation for arbitrarily shaped and arbitrarily oriented particles: a microphysical derivation from statistical electromagnetics, Appl. Opt., 41, 7114–7135.

Mishchenko, M.I., and L.D. Travis, 1997: Satellite retrieval of aerosol properties over the ocean using polarization as well as intensity of reflected sunlight, J. Geophys. Res., 102, 16989–17013.

Mishchenko, M.I., A.A. Lacis, and L.D. Travis, 1994: Errors induced by the neglect of polarization in radiance calculations for Rayleigh-scattering atmospheres, J. Quant. Spectr. Rad. Transfer, 51, 491–510.

Mishchenko M., J. Dlugach, E. Yanovitskij, and N. Zakharova, 1999: Bidirectional reflectance of flat, optically thick particulate layers: an efficient radiative transfer solution and applications to snow and soil surfaces. J. Quant. Spect. Rad. Transfer 63, 409–430.

Mishchenko, M.I., L.D. Travis, and A.A. Lacis, 2002: *Scattering, Absorption, and Emission of Light by Small Particles*, Cambridge: Cambridge University Press.

Mishchenko, M.I., B. Cairns, J.E. Hansen, et al., 2004: Monitoring of aerosol forcing of climate from space: analysis of measurement requirements, J. Quant. Spectr. Rad. Transfer, 88, 149–161.

Mishchenko, M.I., L.D. Travis, and A.A. Lacis, 2006: *Multiple Scattering of Light by Particles: Radiative Transfer and Coherent Backscattering*, Cambridge: Cambridge University Press.

Mishchenko, M.I., B. Cairns, G. Kopp, et al., 2007: Accurate monitoring of terrestrial aerosols and total solar irradiance: introducing the Glory Mission, Bull. Amer. Meteorol. Soc. 88, 677–691.

Monahan, A.H., 2006: The probability distribution of sea surface wind speeds. Part I: Theory and Sea-Winds observations, J. Clim., 19, 497–520.

Moroney, C.R., R. Davies, and J.P. Muller, 2002: Operational retrieval of cloud-top heights using MISR data, IEEE Trans. Geosci. Rem. Sens., 40, 1532–1546.

Muñoz, O., H. Volten, J.F. de Haan, et al., Experimental determination of scattering matrices of randomly oriented fly ash and clay particles at 442 and 633 nm, J. Geophys. Res., 106(D19), 22833–22844, 10.1029/2000JD000164, 2001.

Nadal, F., and F.-M. Bréon, 1999: Parameterization of surface polarized reflectance derived from POLDER spaceborne measurements, IEEE Trans. Geosci. Rem. Sens., 37, 1709–1718.

Nakajima, T., and M. Tanaka, 1988: Algorithms for radiative intensity calculations in moderately thick atmospheres using a truncation approximation, J. Quant. Spectr. Rad. Transfer, 40, 51–69.

Nakajima, T., M. Tanaka, and T. Yamauchi, 1983: Retrieval of the optical properties of aerosols from aureole and extinction data, Appl. Opt., 22, 2951–2959.

Nakajima, T., G. Tonna, R. Rao, et al., 1996: Use of sky brightness measurements from ground for remote sensing of particulate polydispersions, Appl. Opt., 35, 2672–2686.

North, P.R.J., S.A. Briggs, S.E. Plummer, and J.J. Settle, 1999: Retrieval of land surface bidirectional reflectance and aerosol opacity from ATSR-2 multi-angle imagery, IEEE Trans. Geosci. Rem. Sens., 37, 526–537.

North, P.R.J., 2002: Estimation of aerosol opacity and land surface bidirectional reflectance from ATSR-2 dual-angle imagery: operational method and validation, J. Geophys. Res., 107, D4149, doi: 10.1029/2000JD000207.

Nussenzveig, H.M., 1992: *Diffraction Effects in Semiclassical Scattering*, Cambridge: Cambridge University Press.

Nussenzveig, H.M., and W.J. Wiscombe, 1980: Efficiency factors in Mie scattering, Phys. Rev. Lett., 45, 1490–1494.

Okada, K., J. Heintzenberg, K. Kai, Y. Qin, 2001: Shape of atmospheric mineral particles collected in three Chinese arid-regions, Geophys. Res. Lett., 28, 3123–3126, 10.1029/2000GL012798.

O'Neill, N.T., and J.R. Miller, 1984: Combined solar aureole and solar beam extinction measurements, 2, Studies of inferred aerosol size distribution, Appl. Opt., 23, 3697–3704.

Pao, Y.H., 1977: *Optoacoustic Spectroscopy and Detection*, New York: Academic Press.

Perrin, F., 1942: Polarization of light scattered by isotropic opalescent media, J. Chem. Phys., 10, 415–427.

Peterson, J.T., and C.E. Junge, 1971: In: *Man's Impact on Climate, W.H. Matthews, et al. (eds), Cambridge: MIT Press.*

Phillips, B.L., 1962: A technique for numerical solution of certain integral equation of first kind, J. Assoc. Comput. Mach., 9, 84–97.

Rayleigh, Lord, 1871: On the light from the sky, its polarization and colour, Philos. Mag., 41, 107–120, 274–279.

Remer, L.A., Y.J. Kaufman, D. Tanre, et al., 2005: The MODIS aerosol algorithm, products and validation. J. Atmos. Sci., 62, 947–973.

Ricchiazzi P, S. Yang, C. Gautier, D. Sowle, 1998: SBDART: A Research and Teaching Software Tool for Plane-Parallel Radiative Transfer in the Earth's Atmosphere, Bull. Am. Meteorol. Soc., 79, 2101–2114.

Rodgers, C.D., 1976: Retrieval of atmospheric temperature and composition from remote measurements of thermal radiation, Rev. Geophys., 14, 609–624.

Rodgers, C.D., 2000: *Inverse Methods for Atmospheric Sounding: Theory and Practice* (Series on Atmospheric Oceanic and Planetary Physics, vol. 2), Singapore: World Scientific.

Roth, R.S., ed., 1986: *The Bellman Continuum: A Collection of the Works of Richard E. Bellman*, New York: Academic Press.

Rozanov, V.V., 2006: Adjoint radiative transfer equation and inverse problems, in *Light Scattering Reviews*, vol 1, A.A. Kokhanovsky (ed.), Chichester: Springer–Praxis, 339–392.

Rozanov, V.V., and A.A. Kokhanovsky, 2006: The solution of the vector radiative transfer equation using the discrete ordinates technique: selected applications, Atmos. Res., 79, 3–4, 241–265.

Rozenberg, G.V., 1955: Stokes vector-parameter, Uspekhi Fiz. Nauk, 56, 77–110.

Santer, R., V. Carrere, P. Dubuisson, and J.C. Roger, 1999: Atmospheric corrections over land for MERIS, Int. J. Rem. Sens., 20, 1819–1840.

Santer, R., et al., 2000: Atmospheric product over land for MERIS level 2, MERIS Algorithm Theoretical Basis Document, ATBD 2.15, ESA.

Seinfeld, J.H., and S.N. Pandis, 1998: *Atmospheric Chemistry and Physics*, New York: Wiley.

Shettle, E.P., and R.W. Fenn, 1979: Models of aerosols of lower troposphere and the effect of humidity variations on their optical properties, AFCRL Tech. Rep. 79 0214, 100 pp., Air Force Cambridge Res. Lab., Hanscom, Air Force Base, MA.

Shifrin, K.S., 1951: *Light Scattering in a Turbid Medium*, Leningrad: Gostekhteorizdat.

Shifrin, K.S., 2003: Analytical inverse methods fro aerosol retrieval, in *Exploring the Atmosphere by Remote Sensing Techniques*, R. Guzzi (ed.), Berlin: Springer-Verlag, 183–224.

Shiobara, M., and S. Asano, 1994: Estimation of cirrus thickness from sun photometer measurements. J. Appl. Meteor., 33, 672–681.

Siewert, C.E., 1997: On the phase matrix basic law to the scattering of polarized light, Astron. Astrophysics, 109, 195–200.

Siewert, C.E., 2000: A discrete-ordinates solution for radiative-transfer models that include polarization effects, J. Quant. Spectr. Rad. Transfer, 64, 227–254.

Smirnov, A., B.N. Holben, Y.J. Kaufman, O. Dubovik, T.F. Eck, I. Slutsker, C. Pietras, and R. Halthore, 2002: Optical properties of atmospheric aerosol in maritime environments, *J. Atm. Sci.*, 59, 501–523.

Sobolev, V.V., 1956: *Radiative Transfer in Stellar and Planetary Atmospheres*, Moscow: Gostekhteorizdat.

Sobolev V.V., 1975: *Light Scattering in Planetary Atmospheres*, Oxford: Pergamon Press.

Stamnes K., S-Chee Tsay, W. Wiscombe, and K. Jayaweera, 1998: Numerically stable algorithm for discrete-ordinate-method radiative transfer in multiple scattering and emitting layered media, Appl. Opt., 27, 2502–2509.

Stokes, G.G., 1852: On the composition and resolution of streams of polarized light from different sources, Trans. Camb. Phil. Soc., 9, 339–416.

Sugimoto, N., and C.H. Lee, 2006: Characteristics of dust aerosols inferred from lidar depolarization measurements at two wavelengths, Appl. Opt., 45, 7468–7474.

Tarantola, A., 1987: *Inverse Problem Theory: Methods for Data Fitting and Model Parameter Estimation*, New York: Elsevier.

Thomas, G.E., and K. Stamnes, 1999: Radiative transfer in the atmosphere and ocean, N.Y.: Cambridge University Press.

Thomas, G.E., S.M. Dean, E. Carboni, et al., 2005: ATSR-2/AATSR algorithm theoretical base document, ESA Globaerosol ATBD.

Tikhonov, A.N., 1963: On the solution of incorrectly stated problems and a method of regularization, Dokl. Akad. Nauk, 151, 501–504.

Tikhonov, A.N., and V.Y. Arsenin, 1977: *Solution of Ill-Posed Problems*, New York: Wiley.

Tsigaridis, K., et al., 2006: Change in global aerosol composition since preindustrial times, Atmos. Chem. Phys., 6, 5143–5162.

Turchin, V.F., V.P. Kozlov, and M.S. Malkevich, 1970: The use of the methods of mathematical statistics for the solution of the incorrect problems, Usp. Fiz. Nauk, 102, 345–386.

Twomey, S., 1963: On the numerical solution of Fredholm integral equations of the first kind by the inversion of the linear system produced by quadrature, J. Assoc. Comput. Mach., 10, 97–101.

Twomey, S., 1977: *Introduction to the Mathematics of Inversion in Remote Sensing and Indirect Measurements*, New York: Elsevier.

Tynes, H.H., G.W. Kattawar, E.P. Zege, et al., 2001: Monte Carlo and multicomponent approximation methods for vector radiative transfer by use of effective Mueller matrix calculations, Appl.Opt., 40, 400–412.

van de Hulst, H.C., 1957: *Light Scattering by Small Particles*, New York: Wiley.

van de Hulst, H.C., 1980: *Multiple Light Scattering: Tables, Formulas and Applications*, New York: Academic Press.

Veefkind, J.P., G. de Leeuw, and P.A. Durkee, 1998: Retrieval of aerosol optical depth over land using two-angle view satellite radiometry during TARFOX, Geophys. Res. Lett., 25, 3135–3138.

Veefkind, J.P., G. de Leeuw, P. Stammes, and R.B.A. Koelemeijer, 2000: Regional distribution of aerosol over land, derived from ATSR-2 and GOME, Rem. Sens. Environ., 74, 577–386.

Vermote, E.F., D. Tanre, J.L. Deuze, et al., 1997: Second simulation of the satellite signal in the solar spectrum, 6S: An overview, IEEE Trans. Geosci. Rem. Sens., 35, 675–686.

Volten, H., 2001: *Light Scattering by Small Planetary Particles, PhD thesis, Free University of Amsterdam.*

von Hoyningen-Huene, W., et al., 1999: Radiative properties of desert dust and its effect on radiative balance, J. Aerosol Sci., 4, 489–502.

von Hoyningen-Huene, W., and P. Posse, 1997: Nonsphericity of aerosol particles and their contribution to radiative forcing. J. Quant. Spectr. Rad. Transfer 75, 651–668.

von Hoyningen-Huene, W., M. Freitag, J.B. Burrows, 2003: Retrieval of aerosol optical thickness over land surfaces from top-of-atmosphere radiance. J. Geophys. Res., 108, 4260, doi:10.1029/2001JD002018.

Vouk, V., 1948: Projected area of convex bodies, Nature, 162, 330–331.

Wang, M., and H.R. Gordon, 1993: Retrieval of the columnar aerosol phase function and single scattering albedo from sky radiance over the ocean: Simulations, Appl. Opt., 32, 4598–4609.

Wang, M., and H.R. Gordon, 1994: Estimating aerosol optical properties over the oceans with the multi-angle imaging spectroradiometer: some preliminary studies, Appl. Opt., 33, 4042–4057.

Wauben W.M.F., 1992: *Multiple Scattering of Polarized Radiation in Planetary Atmospheres*, PhD thesis, Free University of Amsterdam.

WCP-112 1986: *A Preliminary Cloudless Standard Atmosphere for Radiation Computation*. Geneva: World Meteorological Organization.

Wendisch, M., and W. von Hoyningen-Huene, 1994: Possibility of refractive index determination of atmospheric aerosol particles by groundbased solar extinction and scattering measurements, Atmos. Environ., 28, 785–792.

Whitby, K.T., 1978: The physical characteristics of sulfur aerosols, Atmos. Environ., 12, 135–159.

Winker, D., and B. Hunt, 2006: First results from CALIOP, *Proc. of Amer. Meteorol. Soc. Conference* (http://ams.confex.com/ams/pdfpapers/121113.pdf).

Wittmaack, K., H. Wehnes, U. Heinzmann, and R. Agerer, 2005: An overview on bioaerosols. viewed by scanning electron microscopy, Science of The Total Environment, 346, 1–3, 15, 244–255.

Yanovitskij, E.G., 1997: *Light Scattering in Inhomogeneous Atmospheres*, Berlin: Springer.

Zege, E.P. , and A.A. Kokhanovsky, 1994: Analytical solution to the optical transfer function of a scattering medium with large particles, Appl. Opt., 33, 6547–6554.

Zege, E.P., I.L. Katsev, and A.P. Ivanov, 1991: *Image Transfer through Scattering Media*, Berlin: Springer.

Zhao, F., Z. Gong, H. Hu, et al., 1997: Simultaneous determination of the aerosol complex index of refraction and size distribution from scattering measurements of polarized light, Appl. Opt., 36, 7992–8001.

Index

Printing: Mercedes-Druck, Berlin
Binding: Stein+Lehmann, Berlin

RETURN TO: PHYSICS-ASTRONOMY LIBRARY
351 LeConte Hall   510-642-3122